MW00583240

About Island Press

Since 1984, the nonprofit organization Island Press has been stimulating, shaping, and communicating ideas that are essential for solving environmental problems worldwide. With more than 1,000 titles in print and some 30 new releases each year, we are the nation's leading publisher on environmental issues. We identify innovative thinkers and emerging trends in the environmental field. We work with world-renowned experts and authors to develop cross-disciplinary solutions to environmental challenges.

Island Press designs and executes educational campaigns, in conjunction with our authors, to communicate their critical messages in print, in person, and online using the latest technologies, innovative programs, and the media. Our goal is to reach targeted audiences—scientists, policy makers, environmental advocates, urban planners, the media, and concerned citizens—with information that can be used to create the framework for long-term ecological health and human well-being.

Island Press gratefully acknowledges major support from The Bobolink Foundation, Caldera Foundation, The Curtis and Edith Munson Foundation, The Forrest C. and Frances H. Lattner Foundation, The JPB Foundation, The Kresge Foundation, The Summit Charitable Foundation, Inc., and many other generous organizations and individuals.

Generous support for this publication was provided by Margot and John Ernst.

The opinions expressed in this book are those of the author(s) and do not necessarily reflect the views of our supporters.

White Pine

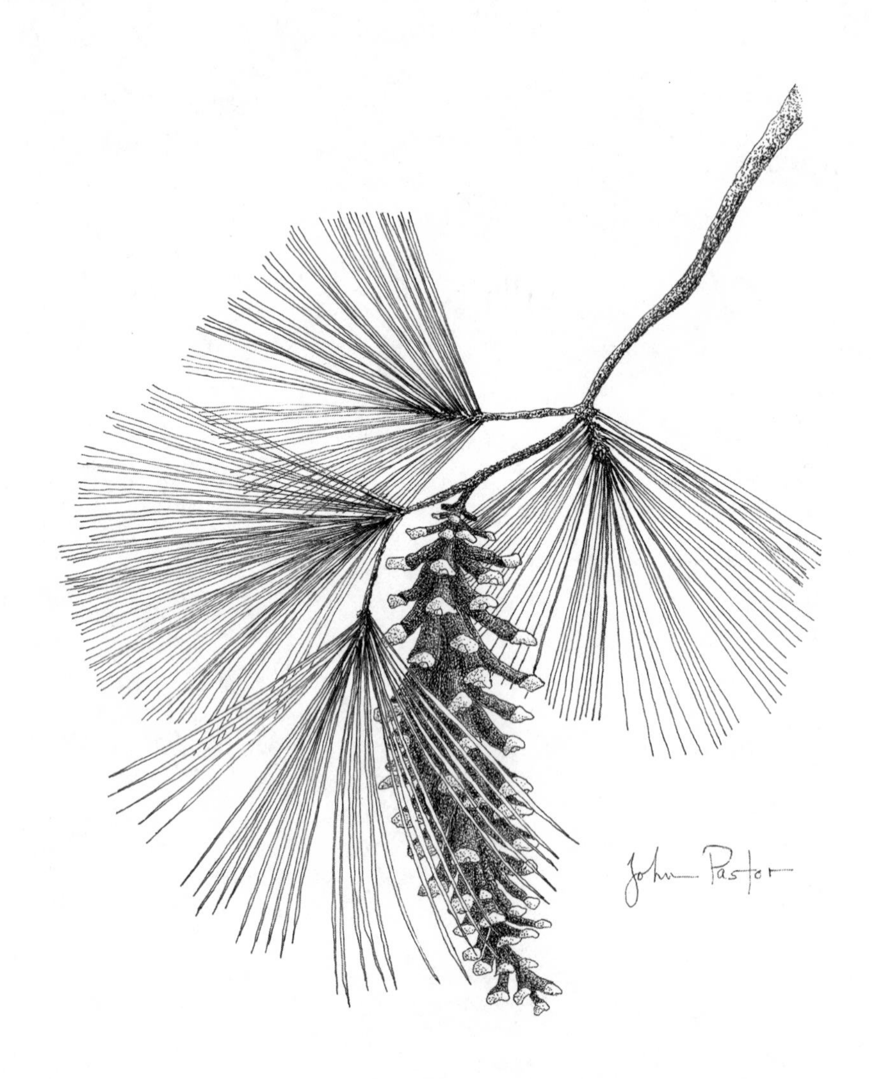
John Pastor

White Pine

THE NATURAL AND HUMAN HISTORY OF A FOUNDATIONAL AMERICAN TREE

John Pastor

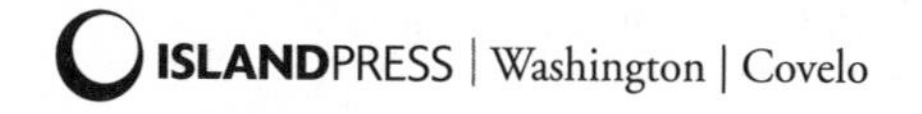

Library of Congress Control Number: 2022934387

All Island Press books are printed on environmentally responsible materials.

Manufactured in the United States of America
10 9 8 7 6 5 4 3 2 1

Keywords: Adirondack Forest Preserve, Algonquin, American Revolution, Bernard Fernow, blister rust, climate change, Civilian Conservation Corps, eastern white pine, ecology, evolution, fire ecology, forest, forest ecology, foundation species, Franklin Delano Roosevelt, George Perkins Marsh, George Washington National Forest, Gifford Pinchot, Henry David Thoreau, Hudson River School, Iroquois, John Muir, logging, lumber, Maine, Minnesota, mycorrhizae, natural history, Nature Conservancy, North Woods, red squirrel, resources management, silviculture, Smokey Bear, spruce, timber, United States Forest Service, Volney Spaulding, watershed, white pine, Wisconsin

For my granddaughter,
Linnea Pastor

Between every two pine trees there is a door leading to a new way of life.

John Muir

Contents

Introduction

My father was a carpenter and a cabinetmaker. Like any good cabinetmaker, he could distinguish different species of wood just from the smell and texture of sawdust in the shop or on the jobsite. The most abundant sawdust in his shop (and there was a lot of it) was from white pine. My father carried the resinous scent of white pine sawdust with him. We had white pine furniture in our living room, slept in white pine beds, and placed our clothes into white pine bureaus. After he returned home from World War II, he framed our house with white pine two-by-fours and paneled the walls in the living and dining rooms with tongue-and-groove pine planks, which he got cheap at the lumberyard because everyone else had switched to using drywall. In my childhood, these walls were a creamy white with a few sepia knots where the branches had once emerged from the trunks. Over the years, the planks darkened so slowly that I didn't notice the change until I moved away to college and then to grad school and beyond. Each time I went home to visit, I noticed that the panels were slowly and steadily turning a rich burnt sienna, like pumpkin pie.

Later, I and my friends and colleagues worked in primeval forests on Blackhawk Island in the Wisconsin River.[1] One stand contained large old-growth white pines. The huge trunks of these trees, between two and three feet in diameter, were free of branches for sixty feet or more above the forest floor. Each pine could easily have provided three clear sixteen-foot logs, prime timber for the region's sawmills. But most likely, the loggers who cut the vast pineries of central and northern Wisconsin in the 1800s did not bother with this particular stand, because it was too small. In addition, floating the pines down the Wisconsin River through the canyons and falls of the Wisconsin Dells immediately downstream just wasn't worth the effort or the danger of logjams. And so, these pines have remained untouched to this day.

Every time we walked into this white pine stand, I felt as if we'd entered a different world. The air was cool and fragrant. The crowns of the pines towered above us and above sugar maples in the understory. During autumn, the maple leaves glowed a scarlet red made even more vivid by their juxtaposition with the deep green pine needles above them. Over the centuries, the dead pine needles and sugar maple leaves that fell to the forest floor decayed and built a black humus that smelled faintly sweet and resinous, like my father's workshop. The images and smells of these big white pines over a maple understory in autumn remain for me the essence of what *forest* means.

I am happy that the loggers spared this stand and regret that more old-growth stands were not so spared. I continue to enjoy hiking and canoeing through remaining old-growth pine stands in wilderness areas around Lake Superior and elsewhere. At the same time, I admire the courage and skill of the loggers who felled the pines from Maine to Minnesota and transported them via oxen, horses, and rivers to the mills. And I obviously have fond remembrances of the look, feel, and smell of the paneling and furniture my father made from the beautiful white pine lumber milled from trees not unlike those that grow in this stand.

Such conflicting thoughts—whether to preserve or harvest, to enjoy the landscape or a product—are not unique to me personally. These conflicting thoughts underpin the challenges of making environmental policies that both conserve the beauty of our landscape and guide sustainable harvest of natural resources to support ourselves. White pine is a model species to help guide our thinking about these problems. Throughout this book, we'll trace how white pine helped shape American environmental thought, from both Native American and colonial perspectives before and during the American Revolution, through the era of clear-cutting, to attempts to illuminate the details of its complex natural history, and finally to historic and current efforts to preserve and restore it. Even in areas where no white pines are found, how we think about the landscape we live in has been shaped in part by our interactions with white pine over the past several centuries.

Eastern white pine (*Pinus strobus*) is a spectacular species, the largest tree in the North Woods, the one that towers over all others. The North Woods is an immense forest encompassing nearly a million square miles and stretching from the shores of Newfoundland to the prairies of western Minnesota; it is bordered by the eastern hardwood forest to the south and the boreal forest to the north. The range of white pine nearly completely overlaps with the territory of the North Woods, where it can be found in pure stands or, more commonly, in mixed stands of sugar maple, yellow birch, beech, red oak, or hemlock. White pines over forty inches in diameter and 150 feet tall were not uncommon in forests from Maine to Minnesota before the logging era. Trees of such sizes can occasionally be found today in protected wilderness areas, although trees one hundred feet tall and two to three feet in diameter are more common. Under optimal conditions and without crown fires, white pine can live to be three hundred years old or greater.[2]

By virtue of its abundance, large size, and long life span, white pine dominates the North Woods and underpins its stability and resilience to fire and storms. Along with hemlock and sugar maple, white pine is what ecologists call a foundation species of the North Woods.[3] Foundation species are widespread plants of large sizes that exert considerable control over the transfers of energy and nutrients throughout the food web.[4] They also structure food webs by providing physical habitat for animals, plants, and fungi and control the microclimate that, in turn, modifies the metabolic rates of animals and plants that coexist with them. A foundation species comprises many individual plants that interact with the food web around them in different ways at different stages of their life cycles. Outside of the North Woods, Douglas fir, redwood, saguaro cactus, big bluestem, and kelp are examples of foundation species in their own ecosystems.

Rare and uncommon species, rather than foundation species, have been the primary focus of conservation biologists. Because foundation species are abundant, we often take them for granted and are not usually concerned about their conservation. A rare species, however, doesn't have the biomass to control and sustain a food web the way a foundation species can, and there are usually only one or two foundation species at the base of most food webs. When they disappear, the food webs they support can rapidly go awry. By the time we recognize that a foundation species is diminishing, it may be too late to do anything about it. Climate change, increased harvesting, the spread of exotic insects and diseases, and many other factors are together threatening numerous foundation species, lending an urgency to their conservation.

White pine has also been a foundation of the cultures of the North Woods, both of Native Americans and European settlers, for the past four hundred years. The Algonquin peoples recognized connections between white pines and other species that they revered, such as bald eagles, a view not unlike the ecological concept of a foundation species. Similarly, white pine played an outsized role in the development of the American character. As the botanist and author Donald Culross Peattie wrote, "No other tree has played so great a role in the life and history of American people."[5] White pine was used to build the ships and houses, barns, and bridges of America from the 1600s through the Civil War and beyond. Before and during the Revolutionary War, white pine was also a symbol of American independence from the king of England: a white pine image adorned many Revolutionary War flags, including the one raised at Bunker Hill. To these utilitarian and patriotic values, Henry David Thoreau added the value of wilderness after he saw what the loss of white pine did to Maine's North Woods. He also began the first scientific study of white pine seedlings and the role white pine plays in the recovery of forests from logging.

The loss of mature forests to white pine logging in the eighteenth and nineteenth centuries helped spark a significant transition in the nation's view of the relationship between man and nature. The destruction of forests was a crucial theme in George Perkins Marsh's development of the idea of watershed protection. Influenced by both Thoreau and Marsh, the Hudson River School of painters created huge canvases of both wild and denuded forest landscapes that also became a force for conservation. White pine (or, as we shall see, its absence) features prominently in some of these paintings. The ideas of Thoreau, Marsh, and the Hudson River School led directly to the establishment of the Adirondack Forest Preserve, the first and still the largest wilderness preserve east of the Rocky Mountains and the location of some of the last uncut white pine forests in the Northeast. The establishment of other wilderness preserves, many also the last remaining strongholds of uncut white pine, followed shortly thereafter.

By the beginning of the twentieth century, Volney Spalding, professor of botany at the University of Michigan; Bernhard Fernow, head of a federal agency that was the precursor of the Forest Service; and Gifford Pinchot, the first chief forester of the US Forest Service, were establishing the scientific basis of white pine's natural history and management.

A few decades later, the planting of millions of white pine seedlings on cutover land by the Civilian Conservation Corps (CCC) helped the North Woods landscape recover from indiscriminate logging. Even more important, the restoration of white pine and other trees helped the young men of the CCC regain a sense of self-worth and helped the nation recover socially from the Great Depression.

These shifts in how we viewed and interacted with white pine reflect the contradictions that have made conservation such a lasting challenge. We love our Paul Bunyan heritage and the beauty of white pine lumber, but we get that wood by harvesting large trees, exactly the sort many people want to preserve. Enormous fires after logging of both eastern and western white pine (a close relative) in the late nineteenth and early twentieth centuries led to Smokey Bear and the Forest Service's "no burn" policy, yet research by a Forest Service ecologist in the 1960s showed that natural white pine forests require a complex fire regime to sustain them. With the demand to protect homes and cabins from wildfires, that complex disturbance regime has to be created by harvesting timber. Today, everyone hopes that the cabin Grandpa built deep in the North Woods beneath the big pines will be in their family for generations to come, but many don't know (or still don't believe) that the warming climate may threaten the survival of the white pines surrounding that cabin. All these threats to and demands of white pine are happening simultaneously, and each amplifies the effect of the others.

Today, government foresters, landowners, sawmill owners, and conservation organizations such as the Nature Conservancy are discovering new ways to manage white pine that mimic natural processes more closely. It is hoped that these approaches will sustain not only white pine but also other species, watersheds, and landscapes that depend on this foundation species of the North Woods. These approaches recognize that a white pine forest is not simply a warehouse of timber but also a protector of watersheds and habitat for game and nongame animals, that fire and other disturbances can both destroy and sustain these forests, and that white pine forests are a foundation of the cultures of the people who live in them.

This is true not only for white pine but for other foundation species and the ecosystems they are a part of as well. By telling the utilitarian, cultural, and scientific stories of white pine, I will explore how a foundation species defines and inspires our views of nature and how we live within a landscape dominated by it. I hope these stories will show how white pine can be a model for the conservation of other foundation species and the ecosystems and people they support.

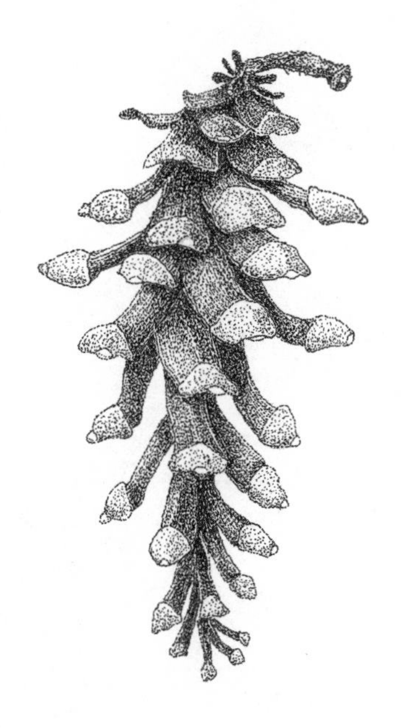

CHAPTER 1

The Evolution and Arrival of White Pine

There is a small old-growth stand of eastern white pine not far from my home in Duluth, Minnesota. It is just off Forest Highway 15 in the Superior National Forest, one of the oldest of our national forests. In the Boundary Waters Canoe Area Wilderness north of this stand, the Superior has many thousands more acres of old-growth white pine forests, but this particular stand encapsulates the history and characteristics of numerous remnant virgin pine forests scattered throughout the North Woods from here to Maine.

I walk into this stand and immediately my eyes travel up and up the straight, clear trunks of colossal trees, pillars touching the sky. Eighty or more feet above me, thick branches extend outward and flare gracefully upward at their tips. The hissing of the needles in the slight breeze makes a soft blanket of sound floating high overhead. A red squirrel perches on one of the branches and screeches sharp, staccato notes. A pileated woodpecker chisels large square holes, twice the size of a playing card, in a massive dead snag; it is foraging for carpenter ants quietly carving their galleries beneath the thick, corrugated, slate-gray bark. Down near the ground, the air enveloping me is soft and scented with needles, resin,

and decaying humus. If smells can have a color, this forest smells deep, deep green. Light pours through gaps in the canopy and pools on the forest floor, where white pine seedlings soak it up. It is May, and bunchberry, twinflower, wintergreen, wild lily of the valley, wood anemone, rose twisted-stalk, and pipsissewa bloom profusely.

In a stand like this, there are many generations of pines, some living and some dead. Several generations of large trees, many of them centuries old, shade seedlings and saplings that have been growing a few years to decades. Some dead snags, perhaps even older than the oldest live trees, still stand. Massive trunks of down trees lie on the forest floor; as they slowly decay, they support soft gardens of mosses, lichens, and balsam fir seedlings. These ancient logs may have been rotting for centuries more than many of the live trees have been growing. Altogether, these generations of white pines may span a thousand years or more.

During the past thousand years, the changing climate of the earth has molded the character of this forest. Each generation of pines was born in a different climate. The trees now rotting into the forest floor lived out most of their lives during the Little Ice Age that ended 150 years ago. The giants whose crowns graze the sky today were seedlings when the Little Ice Age ended in the early 1800s. Then the climate began to warm as the Industrial Revolution, fueled by coal and oil, started spewing carbon dioxide into the atmosphere faster than the world's forests, prairies, and oceans could take it up. Consequently, today's seedlings will mature in a much different climate than that of their parents. During the coming decades, individuals with traits to survive heat and drought stress will transmit the genes for those traits to future generations; individuals without those traits will die.

~

Just as a changing climate over the past thousand years has shaped the development of this white pine stand, so have changes in the earth's climate since the time of the last dinosaurs shaped the evolution of all pines. This particular stand is just one stage in the long journey of white pine to northern Minnesota, spanning over a hundred million years of evolution shaped by a changing climate and shifting tectonic plates.

The best definition of climate I have read is by Kate Marvel, an atmospheric physicist at Columbia University and NASA's Goddard Institute for Space Studies: "a mess of air and water moving together and apart on a rotating sphere."[1] Air and water move together and apart because of the uneven distribution of heat on this rotating sphere we call Earth. Water separates from air when it condenses into raindrops or freezes into snow, sleet, or hail. It recombines with air when it evaporates or sublimates into water vapor. Air and water, in turn, move over the continents and oceans, which exchange water and heat with the atmosphere. The distributions and topographies of continents change as the plates that hold them drift across the earth and crash together to form mountains or separate to form oceans, all the while taking their inhabitants with them.[2] These changes in the continents and oceans, in turn, affect climate.

The interlocking dynamics of continents, oceans, and climate constrain all of life and its evolution. Of particular interest for us, the distribution of water and heat across the earth determines where different plant species can live. The movement of air partly determines the dispersal of pollen and seeds across the land and therefore the migrations of future generations of plants, including trees, across that land. And the changing distribution and topography of the land determine where species can potentially migrate as the climate warms or cools or becomes wetter or drier.

Before we trace the evolutionary journey of pines, especially of white pine, in response to climate change and plate tectonics, we need to review how evolution works. Evolution is the cumulative change in the frequencies of traits in populations (groups of organisms that live in the same area and can breed with one another and produce viable offspring). The frequencies of traits in a population change down through many generations in response to the environment. It is important to remember that individuals do not evolve; populations evolve. Natural selection, a process first theorized by Charles Darwin in *On the Origin of Species*, drives the long-term evolution and diversification of populations and ultimately species. A species is at least one population but more usually a collection of them, some separated from the others by geographic, behavioral, or other barriers yet still able to interbreed if the barriers were to be removed. The area that these populations occupy is known as the species' range.

Natural selection relies on three things: First, individuals vary in the traits that allow them to survive and reproduce. Second, each generation inherits those traits from its parents. Third, some individuals possess traits that allow them to survive longer and pass more of their traits down to their offspring, so the frequency of their traits increases in future generations while the frequencies of traits of other individuals less well adapted to the environment they find themselves in decline. Variability and heritability of traits that result in differential survival and reproduction is natural selection in a nutshell. Individuals of the next generation might mate with other individuals with other traits that also allow them to survive and reproduce better. As a consequence, generation by generation, a population becomes increasingly better adapted to its environment by accumulating traits beneficial to survival and reproduction, so long as the environment doesn't change. If elements of the environment (such as the climate) do change, then other traits may be better for survival and reproduction. The population then evolves in a different direction.

An individual's traits are controlled by its genes, units that are passed down from one generation to the next. Mutation of those genes provides the variability that is the raw material for natural selection to work on. A mutation spreads through the population as the traits it controls are selected. New mutations can spread most rapidly in small, isolated populations where they will not be swamped by the more common genes in a larger population. The longer populations are isolated from one another, the more their traits will diverge. Populations are isolated from one another by many mechanisms, but as we shall shortly see, geographic barriers such as mountain ranges, oceans, ice sheets, and other uninhabitable environments have been particularly important to the evolution of pines.

If and when isolated populations reunite, their gene pools will mix. One of four things can then happen. One of the populations can cause the other to go extinct because it is a better competitor for resources. Or the populations may no longer be able to interbreed because their traits have become so different that successful mating and reproduction are inhibited: they have effectively become two species. Or else, the two populations can hybridize. If the hybrids do not have viable offspring or have fewer offspring than their parents, then their genes will not spread and eventually they will go extinct. In other cases, especially in plants, the hybrids can survive better and produce more descendants than their parents. Those hybrids have become, or are at least on their way to becoming, a new species.

Each of these processes has guided the evolution of pines in response to changing climates and continents during the past 150 million years, resulting in more than 120 species worldwide today.

Before pines evolved, all landmasses were fused into one supercontinent, known as Pangaea. About 175 million years ago, Pangaea began to break apart into huge chunks that eventually became the continents we know today. The southern continents (South America, Africa, Australia, and Antarctica) broke away first, leaving North America and Eurasia still connected in a smaller but still enormous continent known as Laurasia. Then, just before the Jurassic period, North America and Eurasia started separating from each other, unzipping slowly from south to north and opening the Atlantic Ocean. Each continent took its resident plant and animal populations with it, isolating them from populations on other continents and allowing new genes and traits to arise by mutations that were then preserved by natural selection.

The oldest known pine, *Pinus mundayi*, lived 133 to 140 million years ago in what is now Nova Scotia during the Early Cretaceous—a warmer period than ours and long before *Tyrannosaurus rex* flourished.[3] North America and Europe were still connected along their northern fringe, allowing the descendants of this first pine species to spread east into Eurasia and west into North America. By this time, the southern continents were far from North America and Eurasia, which is why pines occur throughout the Northern Hemisphere in North America and Eurasia but are absent in the Southern Hemisphere.[4] Soon thereafter, large basins formed as the final connection between North America and Eurasia was just starting to split. The fossil needles and twigs of *P. mundayi* were encased in the sandstones of the Chaswood Formation that were deposited in these basins.[5]

The fossil needles of *P. mundayi* are grouped two to a bundle, called a fascicle, which is attached to twigs that have resin ducts. On modern pines, the ducts exude terpenes and other flammable resins. The charred fossil twigs of this particular *P. mundayi* suggest that it died in a crown fire, perhaps fueled partly by such resins. It seems that pines were intimately associated with fires from their beginning.[6] In a later chapter of this book, we look at how fires today are still shaping the ecology and evolution of white pine and other pine species.

Approximately twenty million years later, the *Pinus* genus evolved into two main branches, the yellow or "hard" pines and the white or "soft" pines. These two branches of *Pinus* are actually subgenera, designated as *Pinus* and *Strobus*, respectively.[7] The yellow pines include jack pine (*P. banksiana*) and red pine (*P. resinosa*), which currently coexist with white pine in the North Woods. In Eurasia, Scots pine (*P. sylvestris*), another member of the yellow pine branch, is the most abundant pine from the British Isles, through Scandinavia and the Baltic states, and east across all of Siberia. The yellow wood of these pines has a strong turpentine smell from high concentrations of resins. The rather stout and stiff needles of most yellow pines are grouped two or three to a fascicle encased by a small sheath attached to the twig, similar to those of the ancient *P. mundayi*. The sheath is deciduous, and the tree sheds the fascicles with their needles after two to four years.

In contrast to the yellow pines, the wood of the white or "soft" pines has a lighter, almost creamy color. Because the wood of white pines contains less resin than that of yellow pines, it has a softer and more pleasant scent. Modern-day eastern white pine (*P. strobus*) is the defining (or "type") species of this branch, hence the application of the term *strobus* to both the species name of eastern white pine and the subgenus it typifies. Like most of its relatives, eastern white pine has five soft, flexible needles in a fascicle, which falls from the twigs after two to three years. The rubbing of those soft and flexible needles in the breeze creates the gentle whispers that we hear when walking through white pine stands, a sound distinctly different from that produced by the stiffer red and jack pine needles.

The branching of the genus *Pinus* into the yellow and white pines was one of the two most significant events in the evolution of pines. As we shall see throughout this book, the differing susceptibilities of the yellow and white pine branches to diseases and other environmental factors determined their ecology, and the different properties of their wood dictated how the timber industry used them. During the ensuing 120 million years or so since the split between the branches, both the yellow and white pines spread east and west across North America and Eurasia while the continents were still (barely) connected along their northern fringe. Once North America completely separated from Europe, populations and species on each continent evolved into different assemblages of pines.

Beginning about sixty million years ago, at the beginning of the Eocene, the yellow and white pines rapidly radiated into many new species. This was the second important event in the evolution of pines. Constance Millar, a conifer biologist with the US Forest Service, proposed that periodic global swings between warm tropical and cooler temperate climates during the Eocene drove this rapid diversification of pines in North America.[8] Millar's theory of the roles of climate fluctuations has since been verified by the discovery of more fossils and by DNA analyses of modern species, and today it remains the most widely accepted theory of pine biogeography and evolution.[9]

Millar proposed that, when tropical climates were widespread in the Eocene, North America's pine populations contracted into isolated refugia at higher and cooler elevations in the Rocky Mountains. At the time, the Rockies were in the last phases of rising as the now almost completely buried Farallon Plate slid eastward under the westward moving North American Plate.[10] The complex terrain of different mountain ranges and valleys isolated the populations from one another and prevented cross-pollination. Genes mutated in each population. Some of these mutations made the individuals possessing them better adapted to the local environment. Over time, these mutations spread throughout each population.

Later, during relatively cooler temperate climates, the populations expanded out of their mountain refugia onto the plains to the east. Many previously isolated populations could no longer interbreed successfully and had therefore become separate species, while others interbred and formed new hybrids. Some of the hybrids were less able to compete with their parent species and went extinct, but others outcompeted their parent species and drove their parents to extinction or into marginal habitats.

Still later, when the climate entered another warm tropical period, the populations retreated again into mountain refugia. This time, though, the hybrids brought new genes acquired by interbreeding with other populations. Another round of mutation and natural selection in isolated populations followed. Back and forth, back and forth, each expansion and contraction of the populations in response to climate change shuffled their gene pools, selecting for some genes and the traits they control and discarding others across the diversity of microclimates, soils, and habitats in the mountainous terrain. This repeated sifting, winnowing, and recombining of genes of the ancestors of the modern pines made the Rocky Mountains one of the world's centers of pine evolution and diversification.[11] Today, there are forty species of pines in North America. Almost all came from ancestors that spread across the continent after emerging from their Rocky Mountain refugia.

Seven million or so years ago, the climate began a long period of cooling, leading eventually to the formation, expansions, and retreats of the great Pleistocene continental ice sheets. The ice sheets extended from the Arctic southward to the Mid-Atlantic and Midwestern states. The cold episodes of glacial advances during the Pleistocene forced pines once again into refugia, only this time, the refugia were in more southerly temperate climates rather than in cooler high-elevation climates in the Rockies. The expansions, contractions and fragmentations, and reunification of the pine populations were now in a more north–south direction as the ice sheets waxed and waned rather than in an east–west direction into and out of the Rockies as in earlier times.[12] Out of these new scramblings of the pine gene pool, the modern eastern white pine, *Pinus strobus*, finally emerged.

The last great Pleistocene ice sheet, known as the Laurentide Ice Sheet, stretched from Long Island through Pennsylvania and south of the Great Lakes, and finally into Wisconsin, Minnesota, Manitoba, and Saskatchewan. The Laurentide Ice Sheet reached its maximum extent approximately eighteen thousand years ago. Soon after that, the climate began to warm, and the margin of the ice sheet began to retreat northward.

As the ice sheet withdrew, a landscape emerged that was pitted with basins filled with water. Every spring, those lakes and ponds collected the pollen that rained upon them from pines and other trees. Pine pollen grains look like squashed footballs with two air sacs attached to them that help keep them aloft in the wind as they travel from one tree to another, often drifting quite a distance. On June 21, 1860, Thoreau wrote in his journal about the film of pine pollen on Walden Pond and nearby lakes and concluded, "As chemists detect the presence of ozone in the atmosphere by exposing it to a delicately prepared paper, so the lakes detect for us the presence of pine pollen in the atmosphere. They are our *pollinometers*."[13] Year after year, spring after spring, a new layer of pollen drifted through the water column into the sediment. By taking cores from the sediments of a lake, painstakingly sifting through them layer by layer, and recording the relative abundance of the pollen from each species, it's possible to reconstruct how the forest surrounding a lake assembled itself as the climate warmed.

Over the past fifty years or so, palynologists (scientists who study plant pollen and spores) have sampled enough lakes that a picture of the spread of species out of their glacial refuges has begun to develop. Margaret Davis, an ecologist at the University of Minnesota, had the brilliant insight that the dates of first appearance of a species in the pollen record of a region's lakes recorded the arrival of each species to that region as it spread out of its glacial refuge. Davis plotted those dates onto a map of eastern North America and then drew lines known as isochrons that connect points with similar dates of first arrival for each species.[14]

Davis realized that the lines depicted the leading edge of each species as it recolonized the land emerging from the grip of glacial climates. During the maximum extent of the Laurentide Ice Sheet, white pine had retreated to the Atlantic coast of North Carolina and Virginia. The actual range of white pine at this time probably extended farther east onto the continental shelf, which was exposed during the full glacial maximum, when much of the earth's water was locked up into the great ice sheets and sea level was lower. The fossil evidence for this is now buried in the ocean sediments, however. About eleven thousand years ago, white pine spread out of its North Carolina–Virginia refugium by two routes.[15] One route took white pines up the Atlantic Seaboard and then forked into two branches somewhere in southeastern New York State and Western Connecticut. One branch continued northeastward into New England and the Maritime Provinces while the other bent northwestward into Quebec and Eastern Ontario, north of the Great Lakes.[16]

Other white pines spread from the North Carolina–Virginia refugium along a route south of the Great Lakes into Michigan and Wisconsin, entering northern Minnesota and far northwestern Ontario seven thousand years ago, and finally reaching the border of the prairies of northwestern Minnesota only two thousand years ago. The Great Lakes formed a barrier that isolated the white pines traveling along the two routes, just as the different valleys and ranges in the cooler Rocky Mountains isolated different populations of the ancestral pines during the climate fluctuations of the Eocene.

The pine populations that migrated along these two routes evolved separately from each other and from the population that remained behind in North Carolina and Virginia. As the pine populations spread northward into colder climates, they evolved adaptations that help the trees survive the shorter summers and the cold and snow of winter, such as buds that break later in spring, seeds that germinate faster, twigs that

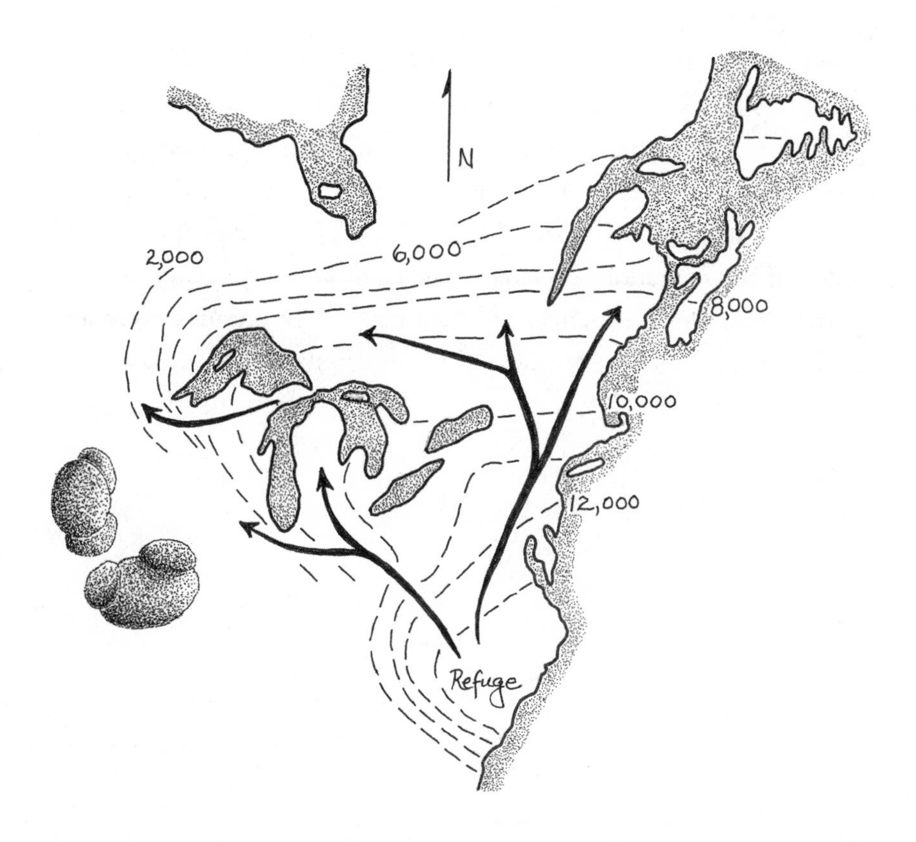

stop growing sooner in fall, and shorter needles that can better survive heavy winter snow.[17]

The descendants of the white pines that left North Carolina and Virginia now tower over me as I walk through the old-growth stand near my home in northern Minnesota, where we began this chapter. I wonder whether the shorter needles of Minnesota pines have different hissing sounds than their longer-needled relatives that remained behind in North Carolina and whether they smell different. It's wonderful to think about how the pines' look, smell, and even sound are the result of shifts and recombinations over millions of years from their origins in the Early Cretaceous to the retreat of the last great ice sheet. Today, the genus *Pinus* is the most species-rich of all conifers. Ever since *Pinus mundayi* first evolved in what is now Nova Scotia, tectonic, climatic, and glacial episodes were the dominant forces controlling the speciation, growth, and geographic distribution of all pines throughout the Northern Hemisphere, white pine included. But humans would become a new force shaping the future of white pine, first as Indigenous peoples followed the migrations of game animals into the deglaciated parts of northeastern North America, and then later, as Europeans arrived and brought their insatiable drive for white pine timber with them.

CHAPTER 2

"A Great Store of Wood and Above All of Pines"

As the ice sheet retreated, the earliest inhabitants of eastern North America, known as Paleo-Indians, followed the migrations of animals such as mastodons, woolly mammoths, moose, and other game species northward to the glacier's edge. At that time, about eleven thousand years ago, the land was occupied by subarctic open spruce woodland. The Paleo-Indians were therefore in place to witness the arrival of white pine from its glacial refuge in Virginia and North Carolina. What followed was a sequence of cultural changes as these Paleo-Indian populations adapted to a pine-dominated landscape.[1] The societies that developed included two large groups of peoples, the Iroquois and Anishinaabeg, both of whom considered white pine to be a sacred tree.

The Iroquois, or Haudenosaunee, began as a confederation of five peoples: the Mohawks, Oneidas, Onandagas, Cayugas, and Senecas.[2] These peoples, who once warred against one another, say that they were united by a prophet named Deganawida the Great Peacemaker. Deganawida, who counted Hiawatha among his followers, called the tribes together into a meeting under a large white pine, the Tree of Long Leaves, and placed an eagle atop the tree to watch over everyone. It is said that Deganawida chose the white pine because each of the five needles in its fascicles represented one of the five peoples.[3]

The Anishinaabeg, a collection of peoples who spoke closely related languages, included the Ojibwa, Potawatomi, Odawa, Mississauga, Nipissing, Menominee, Cree, and Algonquin. Their traditional territories spread nearly across the entire current range of white pine. The Algonquin, in particular, considered white pine (which they called *cigwâtik*) to be the "chief" (*Okima*) of the forest because of its great size. Large white pine monarchs towering above all other trees in the forest had crowns whose unique shapes were signposts that helped orient Algonquin hunting parties as they navigated the landscape. Pine needles, roots, and bark were the base ingredients for many medicines. The white pine was revered partly because it provided important habitat and food to animals also sacred to the Algonquin, such as eagles, moose, deer, and bears, a view not unlike the modern ecological concept of a foundation species. White pine has been called a cultural keystone species for the Algonquin because it played a fundamental role in their medicine, diet, and spiritual practices, and it maintained a supply of other resources, such as game species.[4] And through its role in various stories and ceremonies, white pine also helped maintain the complexity of the tribes' social interactions with the ecology of the land, not unlike the concept of a cultural foundation as presented in the introduction.

The Anishinaabeg peoples, like all Native peoples of North America, did not have any sense of ownership regarding the land in general or white pines or any other species in particular. Indeed, they thought of land not simply as the ground beneath their feet but as something that included all the earth's inhabitants in a great community, a concept not unlike the modern scientific idea of an ecosystem. The traditional ecology of Native peoples was and still is "a cumulative body of knowledge, practice, and belief, evolving by adaptive processes and handed down through generations by culture transmission, about the relationship of living things (including humans) with one another and with their environment."[5]

This perspective is in marked contrast to the much more simplified view of nature as property and a source of commodities that Europeans would bring to the shores of North America. To Europeans, humans were not part of nature; instead, forests existed only to provide products such as white pine lumber. This worldview, as well as new technology that made it possible to harvest large pines, would drive the exploitation of American forests and shape their citizens' relationship to the land for centuries.

~

In the early summer of 1501, three ships sailed westward along the heavy pack of Arctic sea ice flowing out of Davis Strait between Greenland and North America. Their captain, Gaspar Corte-Real, had been sent by King Manuel I of Portugal to investigate the claims of John Cabot, who previously sailed for King Henry VII of England and discovered land somewhere west of the pack ice. Portuguese fishermen had been harvesting cod from the adjacent Grand Banks and Georges Banks for at least a hundred years, and King Manuel thought that any lands bordering these rich fishing grounds were his by right. Corte-Real and his men eventually stumbled upon the wide and deep fjords of the Avalon Peninsula in Newfoundland.[6]

After exploring a few of these fjords, Corte-Real dispatched two of his ships home and continued sailing southwestward along the coast of Newfoundland. Nothing was heard from him again. But the captains and officers of the other two ships, upon arriving in Portugal on October 8, 1501, described the richness of the newly found land to the king. Word quickly spread: by October 19, Pietro Pasqualigo, the Venetian ambassador to Portugal, wrote home to his brothers of a "land rich in berries and . . . a great store of wood and above all of pines for making masts and yards of ships."[7]

These pines were almost certainly white pine, because the only other pine in this region large enough to be used for ship masts, red pine, was found in just two small sites far inland in Newfoundland. Pasquaglio's letter, the first mention of white pines in any European document, was prescient in linking white pine with ships' masts. The transformation of white pine into a commodity would eventually lead to the first revolt against British hegemony of the New World—and eventually to the American Revolution.

Despite the reports of berries and timber, neither the Portuguese nor the British exploited Corte-Real's or Cabot's discoveries immediately. It was not until a century later, on Easter Day 1605, that Captain George Weymouth, under a commission from Baron Thomas Arundell, set sail from Dartmouth Harbor with twenty-nine men aboard the *Archangel*, intending to reach Virginia and establish a colony to be an asylum for English Catholics. The North Atlantic winds were against this southern landfall and forced Weymouth to take a northerly route. After two months at sea, Weymouth piloted his ship through a series of shoals and rocks and spotted first the island of Monhegan and then the "maine land." There he found the mouth of a great river, most likely the fjord of the Penobscot. James Rosier, *Archangel*'s onboard chronicler, wrote that "upon the hilles grow notable high timber trees, masts for ships of 400 tun," a large ship at the time.[8] Masts were made exclusively from pines then, so these "high timber trees" were almost certainly white pines, which tower above all other trees in the coastal forests of New England to this day.

~

Shortly after Rosier and Weymouth wrote about the commercial value of this New World tree, European botanists also made the first scientific description of it. The oldest recorded scientific description of white pine was written by a Royal Professor of Botany and gardener to Queen Mary, Leonard Plukenet, who called it *Pinus Virginiana* in his *Almagestum Botanicum.*[9] Plukenet wrote that he received specimens of *P. virginiana* from John Banister, an Anglican minister and amateur botanist in the Colony of Virginia from 1678 until 1696. Banister was the first botanist to collect and press plant specimens from North America and ship them home to knowledgeable botanists such as Plukenet.[10] Plukenet described the specimens Banister sent as having long, slim cones and very long, thin needles. The only pine from that region of North America that fits this description is what we now call white pine.

But it was Carl Linnaeus, the Swedish botanist who invented the classification system that is still the basis of botany and zoology, who made the first formal taxonomic description of white pine and its relation to other pines. Linnaeus published his description and classification of white pine in a few lines of Latin in his *Species Plantarum.*[11] He based his description not on new specimens sent to him by his pupils but instead mainly on Plukenet's writings and the specimens in Plukenet's herbarium, which he saw during a visit to London in 1736.[12] In his description of white pine, Linnaeus cited Plukenet's work but changed the species name from *virginiana* to *strobus,* from the Greek *strobili,* meaning "long cones." He also added "five needles" to Plukenet's description, which he otherwise copied almost verbatim. Plukenet's private herbarium was bequeathed to his wife upon his death and later acquired by the Bishop of Norwich, from whom it was later purchased by Hans Sloane, who then bequeathed it to the British Museum (Natural History) in London, which he helped to found.[13] Although Plukenet was the first to describe white pine, we recognize Linnaeus as the official designator of white pine as a species because he placed it in the context of a formal taxonomic description of the genus *Pinus.*

It is no coincidence that Pasqualigo, Weymouth, and Rosier all immediately equated white pines with ships' masts. The writings of these early explorers rarely described virgin forests as magnificent or awe-inspiring, as we would today. That view of white pine by European settlers would not be recorded until two centuries later, when Henry David Thoreau took his trips up the Penobscot River into northern Maine. For several centuries after Cabot's voyage, when Western explorers discovered new lands, they saw only commodities to be exploited. These early explorers were no exception, and when they saw these large trees, they envisioned naval stores. And ships' masts were the most valuable of the naval stores because trees of sufficient size and quality to make them had by that time become quite scarce in Europe.

Europeans came upon white pine at a time when they were casting their empires across the world's oceans. Controlling the seas and maritime trade put Britain in an arms race against the other great sea powers: France, the Netherlands, and Spain. To win this race, ships had to be fast and maneuverable, which required acres of sail. In a sailing ship, the force of the wind against the sails was transmitted to the hull and eventually to the bow through the masts. This force was enormous, especially in high winds or as the ship made sharp turns to outflank an enemy in battle. To withstand and transmit these forces, a mast needed to be of considerable size, strength, and suppleness. Great size and strength were needed simply to hold up more than a ton of canvas on the largest ships and withstand the torque of winds at sea. A mast also needed to be supple so it could bend slightly before the wind and then spring back as soon as the wind or the ship changed direction or speed.

Pines were the best—indeed, the only—sort of tree with all three properties. The traits honed by the evolution of pines over millions of years in response to climate change and tree migration suddenly put pine in high demand. Pine was also light and so did not add to the weight of a ship as much as oak or other species did. Before and even for a time after the discovery of white pine, masts for the Royal Navy were made from the yellow *Pinus sylvestris*, known colloquially today as Scots pine.

The *P. sylvestris* trees in Scotland, however, were not large, strong, or flexible enough to match the quality the Royal Navy needed in its masts. The best *P. sylvestris* trees came from the eastern Baltic region in the area around the city of Riga and so were at that time called Riga pines. Riga pines were large, straight, and clear of knots, which would weaken a mast. They also contained large quantities of resins that not only gave pine lumber its characteristic smell but also made the masts supple. The Royal Navy therefore depended heavily on the pine trade from the Baltic region, through the strait of Skagerrak separating Denmark and Norway, and finally into the North Sea to Britain.

As the naval arms race with France, Spain, and the Netherlands accelerated, ships became progressively larger to deploy more and larger cannon. These newer vessels required ever-more sail hoisted on ever-larger masts. A first-class ship of the line with 120 cannon needed a mainmast shaped and cleaned of bark that was at least forty inches in diameter, and bowsprits and foremasts of at least thirty-seven inches. For these masts, the engineering rule was one yard of length for each inch of diameter.[14] The finished mainmast after the harvested tree was formed into the proper shape and taper had to be 120 feet tall. This meant that the tree felled to make the mast must have been on the order of at least forty-five inches in diameter, including the bark, and correspondingly tall.

That is an enormous tree. In any pine forest anywhere, few trees make it to this size. Such trees are rarely found outside of virgin forests, of which precious few remained anywhere in Europe, even at that time. To supply the growing navies of the warring nations, large Riga pines were being harvested faster than they could be replaced, making them scarce by the early 1500s. The Royal Navy then began experimenting with "made masts," which were constructed by splicing smaller trees together.

These made masts did the job, although they were somewhat weaker.

White pines were large enough to make a mast from a single "stick," but the Royal Navy continued to favor Riga pine because the transport distance, and hence the cost, was much less than shipping white pine across the Atlantic. In addition, New England white pines, like all members of the white pine branch of *Pinus*, were less resinous and therefore not so supple as Riga pines of the yellow pine branch. The Royal Navy therefore at first considered white pine an inferior material to Riga pine. It is remarkable that the evolutionary split between the yellow and white pines that happened in the Cretaceous had such a large bearing on the Royal Navy's timber policy a hundred million years later.

Even though the Royal Navy preferred the yellow Riga pine, British colonists knew they had a valuable product in white pine masts and lumber. The first sawmill in America was built in 1623 in York, Maine, less than twenty years after George Weymouth explored the Maine coast and Penobscot Bay.[15] This mill turned sound pines of sufficient size into masts and sawed smaller or defective trees into lumber. As the colonists built hamlets farther inland along rivers and farther east and north into New England, one of the first things they did was build a sawmill. If the British did not want white pine, then the French, Spanish, and Dutch did; the colonists did not think their loyalty to the king meant they should forgo lucrative markets. A large pine suitable for one of the great masts sold then for £100 (upwards of $16,000 in today's currency).[16] Within a few years of the Pilgrims' landing, there was a fortune to be made in white pine.

Pines, however, had little value while still standing in the forest: they attained their maximum value only when hoisting the sails aloft from the deck of a ship. But first the trees had to be delivered to the mills. Felling and moving the behemoth trees was an art practiced by specialists known as mastmen. Once felled, the trees were limbed of branches, many of which were themselves large enough to be milled into lumber. A team of ten to twenty oxen then slid the trimmed trunk to the banks of rivers—first the Piscataqua, which forms the current boundary between New Hampshire and Maine, and later the Penobscot, Kennebec, and other rivers farther east. There, the logs would wait until ice-out, when the spring floods propelled them downstream to the sawmills in Portsmouth, New Hampshire, and later to Falmouth, Maine, and other port cities. Then mastwrights hewed each hundred-plus-foot log into a sixteen-sided prism, which increased the strength of the mast without unduly reducing suppleness.

The finished masts were shipped to Europe in specially made mast ships of four-hundred- and later one-thousand-ton capacity, meaning they could carry forty to one hundred great masts in their cargo holds.[17] At the time, these were some of the largest ships plying the seas, often with a flag bearing a white pine flying atop their own white pine mainmast, a flag that became known as the Flag of New England.

Doing these difficult and often dangerous jobs fostered a great sense of pride among the mastmen, mastwrights, and millwrights. These men were doing jobs on a scale that was only rarely achieved in Europe, and they knew it. That pride, along with the enormous economic value of native white pine on world markets, sparked in the colonists an early sense of independence from the mother country.

In 1652, the economic calculus around white pine suddenly shifted again. During the Anglo-Dutch Wars for control of the high seas, the supply of Riga pines to the Royal Navy came to an abrupt halt when Danish and Dutch navies blockaded the Skagerrak. Deprived of the Riga pines, the Royal Navy had to secure the supply of white pines from New England and prevent their delivery to France, Spain, and the Netherlands.

In 1685, Parliament created the office of Surveyor of Pines and Timber to locate and inventory New England white pines while they still stood in the forests. Then, in 1691, a new Royal Charter to Massachusetts (which included Maine at the time) reserved for the king, William III, "all trees of the diameter of twenty-four inches and upwards at twelve inches from the ground" growing on any land "not heretofore granted by the King to any private person."[18] This was essentially all the land in Massachusetts and Maine that contained any sizable white pine. A series of acts of Parliament soon followed, strengthening these Pine Laws until, in 1721 in the reign of George I, an act of Parliament made it illegal for any "Person or Persons whatsoever . . . to cut, fell or destroy any white Pine Trees, not growing within any Township or the Bounds or Limits thereof, in any of the said Colonies or Plantations, without His Majesty's Royal License for doing so had first been obtained."[19] His Majesty now owned virtually all pines of any commercial value in the colonies.

To enforce the Pine Laws, the Royal Surveyors of Pines and their deputies used broadaxes to blaze an arrow, known as the King's Broad Arrow, on all white pines greater than twenty-four inches in diameter one foot from the ground. This was the first systematic survey of forest resources in North America. But the colonists laughed at these surveyors and their Broad Arrows and cut the trees for their mills anyhow. The colonial trade in mast trees with France and Spain continued without regard for the king's wishes or prerogatives. Trees greater than twenty-four inches that were not made into masts were sawn instead into planks no wider than twenty-three inches; the mill owners could thereby claim that no tree was cut that violated the king's twenty-four-inch rule. Even today, such odd-sized planks can be seen in the floors, siding, and roofs of pre–Revolutionary War homes and barns in New England and the Mid-Atlantic states.

At first, the Royal Surveyors did not strongly enforce the Broad Arrow Laws; indeed, because they knew where the best mast trees were, they even reaped some of the profits from their harvest and export. But when natural resources have high value, corruption soon follows, then as now. This lackadaisical and simultaneously profiteering attitude of the surveyors ended when John Wentworth became governor of New Hampshire in 1766. Wentworth was determined to enforce the Pine Laws. With his blessing, John Sherman, the deputy pine surveyor of New Hampshire, ordered a search of all sawmills in New Hampshire for any pines bearing the Broad Arrow markings. Finding logs with the Broad Arrow in six mills in Goffstown and Weare, Sherman arrested their owners and fined them.

Although the Goffstown mill owners paid the fines, the Weare owners, led by Ebenezer Mudgett, resisted. On April 13, 1772, the sheriff of Hillsborough County, Benjamin Whiting, and his deputy, John Quigley, went to Weare to arrest Mudgett and confiscate the contested pines. Ironically, Whiting and Quigley stayed that night in Weare at the Pine Tree Tavern, now a historic site marked with a large millstone and bronze plaque. In the morning, Mudgett and thirty or forty compatriots descended on the tavern, pulling Whiting and Quigley from their beds and whipping them with pine switches, one lash for each tree Whiting and Quigley confiscated. The sheriff and his deputy were then put on their horses and run out of town through a gauntlet of jeering townspeople. This became known as the Pine Tree Riot, and some historians view the Pine Tree Riot as the precursor to the Boston Tea Party the following year.

~

During the American Revolution, the White Pine Acts and their heavy-handed enforcement made white pine a symbol of the political independence of the colonies. The Massachusetts Minutemen quickly adopted the Flag of New England bearing a white pine emblem and carried it into battle, most notably the Battle of Bunker Hill. This particular flag was later depicted most famously in an oil painting by American artist John Trumbull titled *The Death of General Warren at the Battle of Bunker's Hill, June 17, 1775*. The painting shows a chaos of bayonets, muskets, and red-coated arms raising sabers above the sloping ground, almost all pointing to the white pine on the flag as General Warren lies dying beneath it. Along the opposite edge of the painting and lower than the Flag of New England, a redcoat hoists the flag of the British forces. Trumbull's painting declares the white pine a symbol of righteous insurrection and independence, telling the British that, as far as the Minutemen were concerned, the Broad Arrow markings were null and void and that both white pines and the lives of New Englanders were above the control of the king.

As important as white pine was as a symbol of independence, it also played a substantive role in advancing the American cause during the Revolution. The Royal Navy replaced the masts on each ship of the line of the Royal Navy every three years as they were weakened by dry rot and strain. On January 1, 1775, three months before the Battles of Lexington and Concord, the Royal Navy had only seventy-five masts in reserve that were suitable for mainmasts, foremasts, and bowsprits. Soon after the start of hostilities in 1775, the colonies cut off the supply of masts to Great Britain but not to France and other countries.

Great Britain thought that the war would be over soon and its meager supply of seventy-five spare white pine masts would suffice. As the Revolution dragged on, however, many ships of the Royal Navy were sailing with old and weakened masts while the French ships were well supplied with fresh white pine masts from New England and elsewhere in the rebel colonies. By 1778, the Royal Navy was at a serious disadvantage against the French for control of the seas. In his classic book *Forests and Sea Power*, the historian Robert Albion claims, "The lack of masts [in the Royal Navy] deserves more of a place than it has yet received among the various reasons for England's temporary decline in sea power."[20]

The decline of the Royal Navy came to a head when France signed a Treaty of Alliance with the colonies on February 6, 1778. By April 15, a French fleet under the command of Charles Henri Hector d'Estaing set sail for America to reinforce colonial forces. King George III ordered Admiral John Byron, grandfather of Lord Byron—and Foul-Weather Jack to his sailors—to assemble a fleet to intercept the French and then reinforce British control of the American coastal waters. On July 3, when Byron and his ships were not even halfway across the Atlantic, a strong storm cracked the rotted masts, sending them crashing to the decks.

Ship carpenters tried to splice them back together, but the masts were so weak that the captains had to shorten sail for fear of losing their ships altogether. Byron's squadron was dispersed, with many ships arriving broken and hobbled at American ports from New York to Halifax in late August. As a result, d'Estaing sailed leisurely to America without having had to fire a shot. The Royal Navy never recovered from this disastrous failure of its old white pine masts in time to reinforce and resupply the redcoats. Eventually, the French Navy was able to blockade the mouth of the Chesapeake Bay and prevent supplies and reinforcements from reaching General Cornwallis in Yorktown, Virginia. Washington had only to lay siege on the landward side of Cornwallis's army to force Cornwallis to surrender on October 19, 1781.

With Cornwallis's surrender, the King of England no longer owned the white pines of the former colonies. The Americans now believed that the supply of white pines from New England and New York westward was theirs to cut as they wished in order to build a new nation. This way of thinking not only altered the landscape of the North Woods but dominated Americans' relationship to forests and natural resources for the next 150 years.

CHAPTER 3

A Logger's Paradise

The white pines that supplied the lumber and ships' masts for the American colonies grew mainly in coastal New Hampshire and Maine. But explorations through the wilderness of northern Maine during the Revolutionary War, especially during the Continental Army's march to Quebec City, uncovered much fine white pine within easy reach of the Penobscot and Kennebec Rivers.[1] After the war, the newly free Americans turned northward through Maine, cutting the white pines and driving them down these two rivers to Bangor and other port cities. Then, without hesitation, they cut their way west through the North Woods for a thousand miles, ending in northern Minnesota. Most of this vast land was located in old-growth forests where white pines were mixed with other North Woods species such as maple, red oak, birch, spruce, fir, and, east of Minnesota, hemlock and beech. Although these other species were sometimes harvested to make fine furniture and paneling for the wealthier homes, none were as highly prized as the white pine.

Americans stood on the eastern boundary of what must have seemed a wall of timber a thousand miles thick, with white pine the most valuable species in it. The US Forest Service estimates that as much as six hundred billion board feet of white pine stood in these pre–European settlement forests, and pines greater than thirty inches in diameter and clear of branches for eighty feet or more were not uncommon.[2] This is a stupendous volume of timber, all of it composed of much larger trees and of much greater quality than almost anything left in the British Isles and Europe. Loggers would go far for even a few large white pine per acre such as these.

And none of it belonged to the king anymore. As the European settlers began harvesting white pine and then establishing farms, the Anishinaabeg peoples were driven out of the eastern forests and into the Lake Superior region, where they in turn displaced the various Dakota and Lakota peoples onto the plains.[3] This displacement of Native peoples had tragic consequences that remain with us today. Later, as logging moved into the Lake Superior region, the Anishinaabeg peoples were forced onto reservations as the white settlers and government agents assumed "ownership" of these new lands. The Native peoples who signed the treaties had different views of their relationship to the land than the white settlers, a fact that still underpins much of the conflict over natural resource management today.[4] There is a long history of misunderstanding and abuse of treaty rights by non-Indigenous (mostly white) people in the white pine region of Minnesota and Lake Superior, which began with the harvesting of white pine and continued with the management of other resources such as fish and wild rice.

Because the newly formed American government thought it "owned" the land it occupied after the Revolution, it began to give this tremendous source of wealth away as cheaply and as quickly as possible, a policy that would last for the next century. Lands bearing some of the finest white pine timber sold for around a dollar an acre. In the view of the American government, a nation had to be built; revenue had to be generated; timber was abundant and needed for barns, homes, bridges, and ships; and the land had to be prepared for farmers and merchants.

"Opening up the country" was the common phrase as Americans rushed to harvest white pine and send the timber to growing cities throughout the former colonies as well as the new territories and states westward to the Mississippi. The United States government was in a hurry to settle this land to forestall still extant claims to it by Hudson's Bay Company and France. Eventually, the harvest and milling of white pine in the nineteenth century would add over $4 billion to the nation's economy—three times the revenue generated by California gold and slightly more than the amount added by all mining plus the wheat harvest.[5] The United States was truly built with white pine lumber harvested from the North Woods. But this extraction came at the cost of the cultures and even the lives of the Native peoples who lived in these forests.

The white pine was the first embodiment of Americans' idea that all resources—timber, minerals, water, buffalo, waterfowl, other game birds, deer, and the land—were theirs and no one else's and there was enough to last almost forever. Eighteenth-century Americans thought the government should facilitate rather than regulate the harvest of these resources, a view that persisted for the next century and is espoused by many Americans today. The ideas behind the Homestead Act, the land grants to railroad companies on alternating sides of the tracks across the continent, and almost every other natural resource law enacted during the nineteenth century began with the assumption that, like the white pine, these resources were inexhaustible and "owned" by the American government to be given away to the white settlers.

Well into the 1850s, white pines were still harvested and hewed into masts for the shipbuilding industry in Maine, Massachusetts, and as far south as Maryland.[6] White pine masts held the sails of the American ships of the line in the War of 1812 and then the legendary fleet of clipper ships and whaling ships sailing from New Bedford, Nantucket, and elsewhere. In *Moby-Dick*, Starbuck caws from his perch atop the mainmast, "I am a crow, especially when I stand atop this pine tree here." White pine also decked these ships. When the ships sailed to the tropics, the decking was replaced with teak and the pine was then sold as lumber. But with the rise of ironclad and steam-powered ships after the Civil War, the harvest of white pines for masts ended abruptly as the harvest of pines for lumber increased.

Before 1840, loggers harvested five to seven billion board feet of timber,[7] mostly white pine, from the forests of the southern third of Maine and shipped it downstream to Bangor, the first of the large sawmill towns, where it was milled into lumber rather than masts. As logging expanded northward along the Penobscot and Kennebec Rivers, the supply of white pines declined as they became increasingly remote and where spruce became the dominant conifer. After 1840, the harvest of Maine white pine declined exponentially, and the sizes of white pines delivered to sawmills became smaller and smaller. At the same time, the harvest of spruce, which was scorned by earlier loggers, increased rapidly. By the early 1860s, the harvest of spruce in Maine exceeded that of white pine and has remained dominant ever since.[8]

Thus began a practice that was to be repeated as the timber industry moved region by region westward: a rapid harvest of an unimaginable amount of valuable white pine, followed by an exponential decline and a switch to a wood previously considered inferior. During Thoreau's second trip to Maine in the autumn of 1853, a white pine logger told him that, "what [is] considered a 'tip-top' tree now was not looked at twenty years ago, when he first went into the business; but they succeeded quite well now with what was considered quite inferior timber then."[9] With both great hindsight and foresight in his 1988 paper discussing how these forests became so degraded by this period of wholesale logging, Gordon Baskerville, then dean of the College of Forestry at the University of New Brunswick, termed this practice "lowering of utilization standards," that is, lowering the size, quality, and species that mills would accept.[10]

As the supply of the behemoth white pines greater than thirty inches in diameter declined, the mills began accepting smaller-diameter pines and then inferior species such as spruce.[11] By accepting and milling successively inferior trees, the timber industry thus remained viable in Maine for many decades, but the quality of the trees entering the mills and the lumber leaving it were never the same. Eventually, pulp mills became the dominant market for the small-diameter spruce that sawmills could not turn into lumber, and timber cruisers began to measure logs in cords rather than in board feet. In the meantime, the forests became "degraded," as Baskerville noted, dominated by small, misshapen trees.[12] Almost every extractive industry goes through the same lowering of utilization standards as the resources they exploit are decimated in area after area. For example, the fisheries industry on Georges and Grand Banks off the coast of New England and the Maritime Provinces switched to scallops and lobsters, once considered low-quality catches, as the stocks of cod and halibut crashed because of overfishing.

Even before the spruce harvest overtook that of white pine in Maine, loggers began moving through the Adirondacks, the Catskills, the Mohawk Valley of New York, and the upper reaches of the Susquehanna and Monongahela Rivers in Pennsylvania. By the 1850s, just as the harvest of spruce exceeded that of white pine in Maine, the lumberjacks reached the Lake States of Michigan, Wisconsin, and Minnesota and thought they had entered heaven. Here, they found colossal trees and vast acreages of almost pure white pine on the sandy glacial outwash plains, as well as white pines mixed with northern hardwoods and hemlocks on moraines where there was some silt and clay to hold moisture. This is the country of Hemingway's "Big Two-Hearted River," where his protagonist Nick Adams grows up and where he returns after World War I for solace and to heal.[13]

The supply and quality of white pine just seemed to get better and better as the loggers and timber surveyors moved ever westward and northward.[14] It is no wonder that loggers and the nation continued to think the supply was "inexhaustible" even as their logging frenzy drew nearer to the prairies. By the close of the nineteenth century, 162 billion board feet of white pine were cut in Michigan, 70 billion board feet in Wisconsin, and 36 billion board feet in northeast Minnesota, for a total of 268 billion board feet.[15] This is nearly half the total of 600 billion board feet that the entire North Woods was thought to have harbored before the arrival of European settlers. Between 1875 and 1903, over 7 billion board feet of logs passed through the sawmill town of Stillwater, Minnesota, after being driven down the Saint Croix River separating Minnesota from Wisconsin, exceeding what came from the entire state of Maine.[16] And Stillwater was not even the largest of many sawmill towns on rivers in the region.

The greatest stronghold of white pine in the Lake States was in the valley of the Chippewa River in central Wisconsin. Frederick Weyerhauser, founder of the timber empire that still bears his name, estimated the Chippewa Valley contained 46.6 billion board feet of white pine. Weyerhauser said, "The Chippewa Valley might be called a logger's paradise, a very large part of its area being heavily forested with the finest quality of white pine timber, while rivers, streams, and lakes offered a network of excellent transportation facilities."[17]

Weyerhauser and associates began logging this valley and driving its logs down the Chippewa River to Chippewa Falls. There, this city of sawmills turned many of these raw logs into lumber for the growing farms, hamlets, villages, and cities of the Lake States. The rest were assembled into huge rafts before entering the Mississippi River and being pulled by river tugs as far south as New Orleans. An enormous amount of white pine timber poured out of the Lake States.

How was it possible to cut so much pine almost solely by human and animal muscle power in just a century? It seems incredible that virtually all of America's northeastern forests could be so changed by the removal of this foundation species in a relatively few decades, even as the loggers and lumbermen thought the supply was inexhaustible.

The loggers, teamsters, river drivers, blacksmiths, saw filers, and cooks in the logging camps had great pride in their work, which, along with the cash they gained during winter months, motivated them to return to the lumber camps from their farms after their crops were harvested. Over the years, a seasoned and capable group of loggers, teamsters, and camp personnel developed, increasing the efficiency of the white pine harvest as their experience grew. The loggers' sense of pride led to legends and stories of lumberjacks like Paul Bunyan, who felled the great white pines, becoming part of the culture and lore of the North Woods.[18]

The skilled use of logging technologies became another cultural touchstone for the North Woods. Harvesting a white pine thirty-five to forty-five inches in diameter and weighing as much as eighteen tons has always been extremely dangerous. It required not only courage, skill, and foresight but also new technology to fell and move large masses of wood, a job that could kill you if not done right.

Every task in logging was a problem to be solved partly by human and animal muscle but also by a great deal of ingenuity and intelligence. Felling a large pine required several days of preparation. Smaller, nearby trees first had to be felled and stacked into a cradle, often filled with snow in winter, that would break the fall of the pine and prevent the valuable trunk from shattering into worthless splinters. Once felled and trimmed of their branches, the logs were marked with the company's insignia, then transported to rivers or lakes by oxen pulling sleds across ice roads in winters as cold and hard as an iron hammer. The floodwaters of the spring snowmelts swept rivers of logs downstream to quiet reaches or large holding ponds where they were sorted by ownership and sent into the sawmills around their banks. Many of these sawmills became the nucleus of cities that grew up around them, including Bangor, Maine; Detroit and Saginaw, Michigan; Chippewa Falls, Wisconsin; and Stillwater and Duluth, Minnesota.

Loggers and blacksmiths put a lot of thought into how the axes, saws, and other hand tools could help cut and move logs while minimizing risk. These hand tools were precision-made by blacksmiths who had a deep intuitive grasp of how steel interacted with wood and took great pride in their craft. In no tool was the blacksmith's skill more manifest than the two-man crosscut saw.

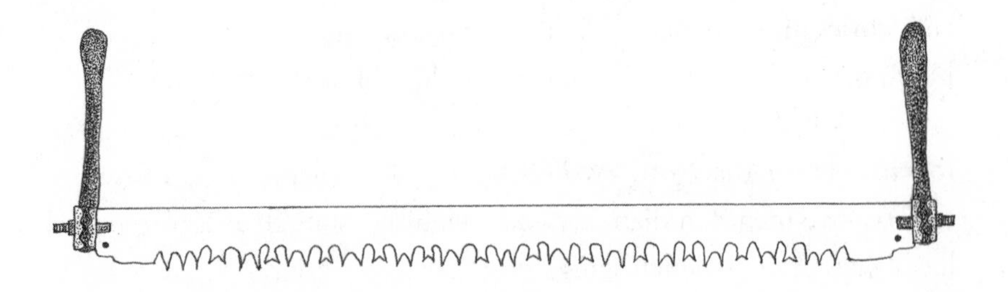

These saws were between six and eight feet long with handles on both ends, each gripped by a sawyer with both hands. The blade of the saw had two kinds of teeth, cutting teeth and raker teeth. The cutting teeth come in pairs that alternate with the single raker teeth. A deep, wide, and round-bottomed gully separates each pair of cutting teeth from the adjacent raker tooth.

As the sawyers pulled the saw back and forth, back and forth through the cut, cutting teeth and raker teeth each performed a different function so they had to be sharpened and set differently. The cutting teeth were in pairs; each tooth of the pair was sharpened to a point and bent, or set, slightly outward in opposite directions so that the points of these teeth inscribed two parallel lines into the wood with each pass of the saw. Each raker tooth had two points sharpened like miniature chisels that were slightly shorter than the cutting teeth and remained vertical in the plane of the blade. As each raker tooth followed the preceding pair of cutting teeth on each sweep, it cut the wood between the two parallel lines inscribed by the preceding cutting teeth and scooped the sawdust and chips into the adjacent round-bottomed gully. As the gully emerged from the log, it dropped the sawdust and chips onto the ground. Almost all logging camps had a filer whose sole job was to sharpen saws and axes. If the saw was not sharpened well, sawing was backbreakingly miserable, giving rise to the name *misery whip*. But when the cutting and raker teeth were set and sharpened properly, the saw bit into the log smoothly; not effortlessly, but with a certain pleasurable rhythm.

~

A few decades into the twentieth century, the white pine timber industry gave one last hurrah. In 1910, in the town of Virginia in northeast Minnesota, Weyerhauser constructed what was then the largest sawmill in the world, milling one million board feet of lumber in a day, a world's record. Just twenty years later, in 1929, the mill closed for good. Although the Chippewa and Superior National Forests in northern Minnesota still supply white pine to local sawmills, the original big trees are almost all gone.

White pine was only one of several seemingly inexhaustible resources that were decimated during the nineteenth century. The range of white pine ends abruptly just thirty miles west of Bemidji, Minnesota, where the great American prairie begins. The American buffalo, which seemed to settlers as inexhaustible as the white pine, roamed these prairies west to the Rockies and south to the Rio Grande. Even as the white pine was being felled to build American homes east of the Mississippi, the buffalo were being slaughtered to provide meat and robes for the wealthier owners of those homes. The near extinction of the buffalo happened almost simultaneously with the depletion of standing white pine timber—a parallel loss that did not go unnoticed, even among lumbermen themselves. It is hard to fathom the simultaneous losses of such abundances as old-growth white pine and buffalo within a century. In his pioneering 1898 study of the white pine industry in Michigan, Wisconsin, and Minnesota, lumber dealer and journalist George Hotchkiss quoted from a US government report: "It will be evident . . . that our virgin coniferous supplies must share the fate which the buffalo has unless a practical application of rational forestry methods . . . is presently inaugurated."[19]

Of course, no such methods would come in time to save most of the original white pine forests, and few white pines remained to seed the next generation. Instead, light seeds from fast-growing species such as aspen parachuted on the wind into newly cut forests, allowing aspen to replace white pine in many of them. The results are still evident today across the North Woods. In the late 1970s, I did my graduate research in an aspen-maple forest in northern Wisconsin that succeeded one of the great pineries cut in the late 1800s. When I worked there, I could still find an occasional boot or broken axle spring buried in the forest floor, left by Scandinavian loggers. Pine stumps were scattered everywhere. Their massive sizes—many of them almost four feet in diameter—kindled thoughts of what the previous, primeval forest must have been like. Even though some of the aspen were as large as thirty inches in diameter and eighty feet tall, it was clear that the current forest was a dwarf compared to the pine forest that had come before. Today, these stumps have rotted into the ground; my generation, or certainly my son's, will be the last to see them.

Pines that were not harvested by the early twentieth century faced another set of challenges: they were often killed by enormous fires that swept through the logging debris. Some loggers and farmers intentionally ignited fires to clear the ground of debris, others were lit by sparks from the smokestacks of steam locomotive engines or from their wheels grinding against the rails, and lightning ignited the rest. The scales and frequencies of these fires are hard to believe. The Peshtigo Fire burned nearly 2,000 square miles in Wisconsin and the Upper Peninsula of Michigan in 1871. Ten years later, the Thumb Fire burned 1,500 square miles of the "thumb" of Michigan's Lower Peninsula, followed in three years by the Hinckley Fire in east-central Minnesota, which burned nearly 400 square miles. In 1910, the Baudette Fire in northwest Minnesota burned 450–550 square miles, and in 1918, the Cloquet Fire burned nearly 400 square miles in northeastern Minnesota. Collectively, these five fires alone burned an area the size of Rhode Island, Connecticut, and Delaware combined. Numerous smaller fires burned an untold additional acreage. Some of these fires were so hot they burned off the topsoil, leaving the mineral soil exposed to eroding rains and spring snowmelt.[20]

A few of these original pine forests did survive both logging and fire. Most timber companies could see that the vast pineries of the North Woods were coming to an end, so many moved out to the Pacific Northwest, where they could log the even larger Douglas fir. By 1905, the number of board feet cut annually in the state of Washington surpassed what was being cut in the entire Lake States of Michigan, Wisconsin, and Minnesota. The departure of the timber companies to the west left isolated stands of uncut white pine, especially in the Lake Superior region. Even today, hikers, hunters, and scientists stumble upon such stands.

This happened to me when colleagues and I were mapping the history of beaver ponds and forests on the Kabetogama Peninsula in Voyageurs National Park near International Falls, Minnesota (the Frostbite Falls of *Rocky and Bullwinkle* fame), using air photos taken at successively earlier decades in the past back to 1928. While these forests are now and always have been mostly aspen, spruce, fir, and birch, there were also isolated patches and pockets of white and red pines. One especially large white pine stand, about two hundred acres, persisted on all photos from 1928 forward. It stood on the north shore of a long fjord of Lake Kabetogama, which had been cut into the Kabetogama Pensinsula by the great Laurentide Ice Sheet eighteen thousand years ago. To the west of the stand, a fresh harvest scar visible on the 1928 photo signaled that the pines there had all been cut. Hundreds, even thousands, of felled trees were corralled in the fjord by a chain, or in loggers' parlance a "boom," of logs across its mouth.

Apparently, this white pine stand remained uncut, abandoned for larger prospects elsewhere. This meant that it might be a primeval forest. At two hundred acres, it would be the largest old-growth white pine stand in Minnesota outside the Boundary Waters Canoe Area Wilderness to the east.

I traveled to this stand to find out—once in it, my jaw dropped. Here were stately white and red pines as far as I could see through the woods. Most of these pines were about two and a half feet in diameter and probably upwards of a century old. The forest floor was thick and springy with needles, moss, and humus and profusely dotted with lesser rattlesnake plantain orchid (*Goodyera repens*), bunchberry (*Cornus canadensis*), wild lily of the valley (*Maianthemum canadense*), and many other flora not seen in such abundance outside old-growth pine forests.

And then, a hundred yards or so ahead, I saw what appeared to be a truly giant white pine towering over the rest. As I drew closer to it, I could feel how its enormous crown dominated its surroundings. Beneath the crown, an accumulation of centuries of needles, bark, and humus rose gradually to the base of the trunk. This tree and the soil it made was an ecosystem in itself. The tree was four feet in diameter at the height of my shoulders; below that, the trunk flared to its base like the bell of a trombone, easily attaining six feet in diameter at the forest floor. This was one of the largest pines I had ever seen, anywhere. As I walked slowly around it, I noticed an old fire scar on the northeast side of the tree, almost completely grown over with bark. None of the other smaller, and apparently younger, trees had fire scars. There had probably not been any fire in this stand for a century or more, and this tree likely was the parent of many of the nearby and slightly smaller white pines. This magnificent tree had survived this fire and probably others before it, along with strong winds, heavy coatings of ice and snow, and other insults during its long life, to say nothing of its lucky escape from the ax and crosscut saw.

I had mixed feelings wandering through the stand on that day and during future visits. On the one hand, I was awestruck, as I always feel when I am in a forest that predates the arrival of Europeans on this continent. It is possible that I was one of the very few, perhaps even the only, person to have set foot in this forest. Maybe a few of the Dakota, Lakota, or Ojibwa who lived in the area before Europeans, a French voyageur, or a timber cruiser surveying the potential for lumber preceded me—or none of these. I felt then, and still feel today, as if I were present at the beginning of the world—or at least the beginning of the pines. I am glad the pines were still there for me to find.

I am sure that many of the loggers who had cut similar pine forests all the way east to Maine must have had the same sense of awe when they first stepped into a stand such as this one. But they had a job to do, and they did it with great skill and courage that I also cannot but help admire. Those lumberjacks did not make the decisions to decimate the white pines. Others who sat in faraway offices made those decisions and became wealthy as a result. Then, without a second thought, they went on to exploit the large western white pines, Douglas firs, and redwoods on the other side of the prairie and became even wealthier. There is nothing wrong with wealth in itself, but to me there is something deeply wrong in acquiring wealth by liquidating natural resources, especially the beautiful monarch pines of the North Woods, as well as forcing Native peoples off the land just for the sake of profit.

There is wealth, however, in landscapes where we can be awed. At the time of the great white pine harvest, few if any of the Americans or Europeans who cleared the forests believed that. But one man did begin to reconsider the conventional value of nature on a trip through the North Woods of Maine. That trip changed his life and shaped the way many Americans now think about forests and wilderness.

CHAPTER 4

Thoreau, the Maine Woods, Forest Succession, and Faith in a Seed

"A pine cut down . . . is no more a pine than a dead human carcass is a man."[1] With these words, Henry David Thoreau offered an alternative view of the value of white pines than the one accepted by the loggers who were busy harvesting and hauling them to the sawmills. In the 1840s and 1850s, Thoreau took three long trips into the Maine woods, spanning the decades when the harvest of spruce in Maine overtook that of white pine.[2] From one trip to the next, Thoreau traveled deeper and deeper into Maine, visited several lumber camps, swooned at the beauty of the North Woods, and became increasingly angry at the destruction of the legendary white pine forests. These trips later inspired him to make some of the first scientific observations of the ecology of white pine. Thoreau's magazine articles about these trips to Maine, posthumously compiled in *The Maine Woods*, as well as his natural history observations and investigations of the ecology of white pine, have not been given the attention they deserve, despite the voluminous scholarship on Thoreau. Thoreau's travels in Maine were a crucial spark for his natural history observations and innovative ideas about the value of wilderness.

In 1846, George Thatcher, who was married to Thoreau's cousin Rebecca Billings, invited Thoreau to accompany him on a trip north along the Penobscot River. That spring, severe floods on the West Branch of the Penobscot had broken logging dams and stranded white pine and spruce logs against its banks. Thatcher, who was in the lumber business in Bangor, wanted to find out if any of this stranded timber was his. Thoreau agreed to go along, perhaps inspired by an earlier trip to Bangor, where he was unable to find a teaching position but heard a Penobscot Indian describe the beautiful country farther north.

Thoreau, Thatcher, and companions left Bangor on September 1, 1846, and began their travels up the Penobscot River Valley by foot and horse and buggy.[3] On the fifth of September, they reached the fork of the West and East Branches of the Penobscot, seventy-five miles upstream of Bangor. There, they switched to bateaux to continue up the East Branch because "there was now no road further, the river being the only highway." For thirty miles, they saw "only half a dozen log huts confined to [the river's] banks."[4]

Eventually, they came upon a logging camp that had been used in some winters past but stood uninhabited now. As Thoreau wrote,

> The camps and the hovel for the cattle were hardly distinguishable, except that the latter had no chimney. . . . These camps were about twenty feet long by fifteen wide, built of . . . logs—hemlock, cedar, spruce, or yellow birch—notched together at the ends, to the height of three or four feet, then of smaller logs resting upon traverse ones at the ends, each of the last successively shorter than the other, to form the roof. . . . The logger's camp is as completely in the woods as a fungus at the foot of a pine in a swamp.[5]

Thus was the extremely primitive nature of the temporary winter quarters for the loggers.

But it appeared that this camp was at the end of its useful life because, as Thoreau wrote, "We saw only the stumps of white pine here, some of them of great size, these having already been culled out, being the only tree much sought after. . . . It was the pine alone, chiefly the white pine, that tempted any but the hunter to precede us on this route."[6]

The way north ended with a climb of Mount Katahdin (or *Ktaadn*, as Thoreau spelled it) before returning by the same route, reaching Bangor by September 10. After his first trip, at the end of the first chapter of *The Maine Woods*, Thoreau penned what is in my opinion the most beautiful description of the North Woods ever written:

> It is a country full of evergreen trees, of mossy silver birches and watery maples, the ground dotted with insipid small red berries, and strewn with damp and moss-grown rocks—a country diversified with innumerable lakes and rapid streams, peopled with trout . . . with salmon, shad and pickerel, and other fishes: the forest resounding at rare intervals with the note of the chicadee, the blue-jay, and the woodpecker, the scream of the fish-hawk and the eagle,[7] the laugh of the loon, and the whistle of ducks along the solitary streams; and, at night, the hooting of owls and howling of wolves . . . the home of the moose, the bear, the caribou, the wolf, the beaver, and the Indian. Who shall describe the inexpressible tenderness and immortal life of the grim forest, where Nature, though it be mid-winter, is ever in her spring, where the moss-grown and decaying trees are not old, but seem to enjoy a perpetual youth . . . ?
>
> What a place to live, what a place die and be buried in![8]

Full of evergreen trees, loons, moose, wolves, caribou, bears, beavers, and the songs of birds, yes. But no longer of white pines.

~

Seven years later, in 1853, Thatcher again invited Thoreau on a trip north into the Maine woods, this time to hunt moose. The party headed northwestward on September 15, traveling sixty miles in an open wagon, catching a steamer boat that sailed the fifteen miles up Moosehead Lake, and portaging to the headwaters of the West Branch of the Penobscot River, down which they paddled by birch bark canoe. Along the way, they shot one moose (a cow) and, on September 18, reached their northernmost position, where the West Branch enters the northern end of Chesuncook Lake. Other than bagging a moose, finding large and uncut virgin white pines was clearly one of Thoreau's major goals for this trip,[9] but he saw only a small grove of young white pine and scattered defective pines left by loggers, noting, "All the rest of the pines had been driven off."[10] This started to frustrate Thoreau: Where were the pines? Could loggers possibly have taken all of them? Was this the best use of this species and these forests? And was this even moral?

Reading Thoreau's journal entries from this time, one can feel how his anger and righteous bile grew within him as he pondered these questions from his canoe. First, he found it "strange that so few ever come into the woods to see how the pine lives and grows and spires, lifting its evergreen arms to the light . . . but are content to behold it in the shape of many broad boards brought to market, and deem *that* its true success!" (Thoreau's italics)

He goes on: "But the pine is no more lumber than man is, and to be made into boards and houses is no more its true and highest use than the truest use of man is to be cut down and made into manure.[11] There is a higher law affecting our relations to pine as well as to men."[12]

By using the phrase *higher law*, Thoreau invoked a theological argument that would have been revolutionary at the time. The concept of higher moral laws was much in the air, especially among the transcendentalists led by Ralph Waldo Emerson, Thoreau's mentor. The phrase has its origins in Roman and Judeo-Christian ethics. In his essay *On the Republic*, Cicero wrote that natural law is higher than any legislative or monarchial decree because "it is in accordance with nature, applies to all men, and is unchangeable and eternal."[13] Cicero thought that obedience to natural law would always summon people to their duty and restrain them from doing wrong. The transcendentalists of Thoreau's time thought that higher laws transcended all legislated laws, including the US Constitution.

It was Ralph Waldo Emerson who linked the legal laws of human societies to the higher natural laws that govern the entire universe. Laura Dassow Walls, biographer of both Emerson and Thoreau, wrote that Emerson believed physical Nature is the embodiment of moral truth.[14] When Thoreau invoked Emerson's higher moral law embodied in Nature to argue against the wholesale removal of white pine throughout the Maine Woods, he applied the same reasoning that the abolitionists used to argue against the removal of Africans from their homelands into slavery. To Thoreau, forcing fellow human beings into slavery and transforming large white pines into profits were both violations of the higher laws of Nature. In his 1849 essay *Civil Disobedience*, Thoreau argued that higher moral laws are the true "fountainhead" of all laws legislated by governments. The philosophical concept of a higher law and its moral implications were on Thoreau's mind in September of 1853 while he paddled a canoe through the North Maine Woods to Lake Chesuncook.

Who, then, Thoreau asked, speaks for the pines? "Is it the lumberman who is the friend and lover of the pine, stands nearest to it, and understands its nature best? . . . No, no, it is the poet; who loves them as his own shadow . . . and lets them stand."[15] The poet, according to

Thoreau, speaks for the pines because he loves them without any vested interest to turn them into commodities for mere profit.

It must have come as a shock to Thoreau not to see any remnant large stands of virgin pine, or even a single large tree, in two long journeys into the heart of the white pine country. At this point, Thoreau turned completely against logging and concluded that the loggers and logging companies had been engaged in a "war against pine" seemingly forgetting that his cabin at Walden was made from "arrowy pines, still in their youth," which he himself had cut.[16] Thoreau ran headlong into a contradiction that remains at the core of today's environmental issues: We need natural resources such as white pine, but how much moral authority do we have to harvest them? When we harvest natural resources, what are our moral obligations and to whom are they owed? To the trees, to the land from which they grow, or to future generations? Or to all three?

But it wasn't just the loss of white pine trees that disturbed Thoreau. He began to glimpse what the loss of white pines meant for the lives of other species, writing, "By his mere presence, almost, [the logger] changes the nature of the trees as no other creature does. . . . [The forest] has lost its wild, damp, and shaggy look, the countless fallen and decaying trees are gone, and consequently that thick coat of moss which lived on them is gone, too."[17] Thoreau was here groping toward a vision of a forest not as a lumberyard but as an integrated community in which white pine performs a set of crucial functions.

But how could we learn about the ecological roles of white pine and other species if the loggers and lumbermen remove them before we get a chance to study them? Thoreau pointed out that, in the community of naturalists and scientists he respected, "We have as yet no adequate account of a primitive pine-forest." He noted that an atlas commonly used in schools limited the woodlands of North America to the deciduous forests of the Ohio River Valley and that "New Brunswick and

Maine are exhibited as bare as Greenland." Ironically, Thoreau wrote, the Scottish botanist John Loudon was then writing that the North American pine forests were the most extensive in the world.[18]

In order for current and future generations to experience and understand the grandeur of white pine forests, Thoreau proposed that if the kings of England could formerly have their own private forests, then "why should not we, who have renounced the king's authority, have our national preserves . . . in which the bear and panther and some even of the hunter race may still exist . . . [where] our forests [exist], not to [merely] hold the king's game . . . [but] for inspiration and our own true recreation?"[19] No better description of the purpose of a public natural preserve for the sake of all the people had ever been made, and this more than a decade before the establishment of Yellowstone, our first official national park.

~

Back in Bangor at the end of this trip, Thoreau had dinner with a relative named Charles Lowell.[20] Lowell told him of a giant white pine cut the previous winter that scaled 4,500 board feet and fetched ninety dollars (nearly $3,000 in today's currency) at a Bangor sawmill. To get to this single massive tree, the loggers cut a road three and a half miles through the forest. Apparently, this pine came from farther north in Maine than Thoreau had yet to travel. This was the land dividing the headwaters of the East Branch of the Penobscot River (up which Thoreau had paddled on his 1846 trip) from those of the Allagash River, which eventually merges with the Saint John River and forms the boundary between Maine and New Brunswick. The fact that loggers found it worthwhile to build a road more than three miles long to cut a single tree should have alerted Thoreau to the scarcity of white pine even in that remote part of Maine. But Lowell assured him that untouched white pine could be found in this region.[21]

And so, in 1857, Thoreau embarked on his longest and last trip into the Maine woods, to the Allagash country far north of Mount Katahdin. On this trip, Thoreau roamed by foot and canoe, guided by a Penobscot man named Joe Polis. The party began traveling by stage up the same route as his 1853 trip to Moosehead Lake and climbed into birch bark canoes to head down the West Branch of the Penobscot River past Chesuncook Lake, where Thoreau's previous trip ended, until they reached what was then known as the Height of Land. This was an extensive plateau in the northern Appalachians that formed a very wet and swampy divide, where the water seemed unable to decide where to drain. Atop the divide lay Chamberlain Lake, out of which flowed both the Saint John–Allagash Rivers north to Canada and the Bay of Fundy and the East Branch of the Penobscot River south to its confluence with the West Branch and thence to Bangor.

Because the water in this region flowed both north to the mills in New Brunswick as well as south to the Bangor mills, logs could be driven either way. To ensure that most of the logs went to the Bangor mills, American lumbermen built a canal and dam, raising water levels so that white pine logs could be floated into the canal and then south along the East Branch of the Penobscot River. This prevented the logs from being floated north along the Saint John–Allagash Rivers to Canada. Thoreau followed this American route by canoe down the Penobscot all the way back to Bangor.

Before embarking on the trip, Thoreau had read the 1836–39 *Reports on the Geology of the State of Maine* by Charles T. Jackson, Maine's first state geologist. In *The Maine Woods*, Thoreau quoted Jackson as saying that this region was the area of "the best pine in the state twenty years ago."[22] That statement seems to have been a motivating factor for Thoreau to take the trip. In addition, timber surveyors in the region also reported white pine to grow in "social habit . . . in communities . . . towering above the surrounding forest, or else they form extensive forests by themselves"; Thoreau "would have liked to come across a large community of pine, which had never been invaded by the lumbering army" but he did not.[23]

Instead, he saw and heard lumberjacks brag about having cut trees whose stumps were large enough to support a team of oxen, to which Thoreau retorted, "Why, my dear sir, the tree might have stood on its own stump, and a great deal more comfortably than . . . a yoke of oxen can, if you had not cut it down."[24] At the head of the East Branch of the Penobscot River, Thoreau reported seeing, "lofty dead white pines slanted halfway over the stream. I saw some monsters there, nearly destitute of branches, and scarcely diminishing in diameter for eighty or ninety feet."[25] Other than the defective pines Thoreau saw in the 1853 trip, these dead pines were the only virgin pines still standing that he saw on his three trips through the Maine woods.

Having confirmed his fears—that northern Maine had been nearly cleared of large white pine, completely altering the landscape—Thoreau turned his attention to the forest as a whole. If he could understand the web of relationships that governed forest life, might there be a way to repair these relationships and heal the forest once logging ended?

~

"The red squirrel should be drawn with a pine cone."[26]

With this entry in his journal, Thoreau began a quest that lasted the final nine years of his life: to understand the connections between white pines, red squirrels, oaks, and the dynamics of forests in and around his home in Concord, Massachusetts. These thoughts and investigations may be the first time any American began to seriously consider how different species in a forest influence and depend on one another. It was a new way of looking at the full complexity of a forest instead of the contemporary lumberman's view of a forest as a warehouse of timber. In Thoreau's journal entries on this topic, we begin

to get a glimpse of white pine as a foundation species of the North Woods. These entries became a common thread through his journal, culminating in his essay "The Succession of Forest Trees," which was one of the first if not the first written analysis of how one tree species replaces another after disturbances, and in his final book, *The Dispersion of Seeds*. In all these investigations, Thoreau was guided by his "great faith in a seed. . . . Convince me that you have a seed there, and I am prepared to expect wonders."[27]

In these writings, Thoreau stressed repeatedly that trees and forests spring up solely from seeds. This is so obvious to us now that we wonder how anyone can think otherwise. But in the 1850s, the idea of spontaneous generation—the generation of living organisms from nonliving matter—was still held to be credible by many intelligent people. Louis Pasteur had just recently demonstrated that bacteria can emerge only from prior generations of bacteria, but when many people at that time saw plants springing from the forest floor, they assumed that some magic vital force transformed decaying leaves into hopeful seedlings. Even Horace Greeley, the editor and publisher of the *New York Tribune*, engaged in a spirited exchange of letters with Thoreau on this point, which he published in his newspaper.[28] Fresh from the disappointment of not seeing old white pine on his trips to Maine, Thoreau grasped that, if we want to understand how these pine forests became established and restore them, then we would have to start with the seeds and follow the tree and its intersections with animals, wind, and soil throughout its life cycle.

Although the original large white pines in Massachusetts had mostly been logged before Thoreau was born, the smaller white pines loggers ignored were now thirty or forty years old and entering the cone-producing stage of their life cycle. Thoreau began his investigations by watching how and when red squirrels gathered those cones for their seeds. In early September, red squirrels cut the green cones that are held in the tops of crowns of white pines. The squirrels throw them to the ground and then, beginning at the base of the cone, dismember them to get the seeds, held in pairs at the base of each scale.[29]

As Thoreau observed, "Every white pine cone that falls to earth [in autumn] before opening and losing its seeds . . . is cut off by a squirrel, and they begin to pluck them long before they are ripe. . . . Their design . . . in cutting them off green is, partly, to prevent their opening and losing their seeds, for these are the ones for which they dig through the snow, and the only white pine cones which contain anything then."[30]

The cones that red squirrels do not cut remain high in the crowns. When they open "in the sun and wind" later in September and October, "away go the seeds of future forests flying far and wide."[31] On a number of occasions, Thoreau observed white pines spreading over hillsides and pastures and along stone walls against which their seeds collided as they were blown by the wind. Often the nearest cone-bearing tree was fifty or sixty rods (16.5 feet) distant.[32] Thoreau observed that pine groves contain one or more large "patriarchs surrounded by their children, while a third generation shows itself yet farther off."[33]

White pines, as well as other species of pines, therefore "infect" open land much like an epidemic, with initial foci of infection (the first established trees that become patriarchs), which spread outward when progeny become established around them and later serve as secondary foci for the third generation.[34] In a natural white pine stand, this cycle often results in groups of trees around the same age spaced about thirty years apart (the length of time it takes for a white pine to bear cones), as opposed to plantations, in which all the trees are of the same age because they were planted at the same time.

On April 28, 1856, George Hubbard, a neighbor, told Thoreau that when white pine stands were cut down, he often saw numerous oaks spring up in their place. This detail interested Thoreau so much that he wrote a memo to himself, "Let me look at the site of some thick pine woods which I remember, and see what has sprung up." He did so a few weeks later, confirming what Hubbard had told him.[35] In Thoreau's mind, these observations raised two questions: How do the acorns get into the pine stands, and why do they survive there better than in their native oak stands?

Enter the red squirrel once again. In September and October of 1857, Thoreau recorded in his journal that he saw red squirrels transporting acorns from oak stands into pine stands, where the squirrels could find cover from predators and the cold at night. There, the squirrels buried the acorns under the pine needles that make up the thick and fluffy forest floor. At this time of year, acorns that naturally fell from their parent trees onto the forest floor had already decayed. The white oak's deciduous crowns had also shed their leaves, so the heat of the day would escape during the night and the ground would be covered with frost by morning. Thoreau thought that these frosts killed the acorns exposed on the forest floor beneath the bare oak crowns. In the pine stand, however, the blanket of dead pine needles on the ground and the still-green pine canopy above would hold in the heat of the day and protect the acorns from early frosts, thus enabling the acorns to survive the fall and winter.[36] Thoreau also suggested that the tannins from decaying oak leaves might somehow inhibit their seedlings from growing, a phenomenon now known as alleopathy.[37]

The red squirrel and the wind, therefore, are two agents that mediate the dispersion of white pines and oaks across the Concord landscape. As Thoreau told the Middlesex Agricultural Society gathering in September 1860, "While the wind is conveying the seeds of pines into hard woods and open lands, the squirrels and other animals are conveying the seeds of oaks into the pine woods, and thus a rotation of crops is kept up."[38] Thoreau later noted, "It has long been known to observers that squirrels bury nuts in the ground, but I am not aware that anyone has thus accounted for the regular succession of forests."[39] Many people, then and now, have seen both windblown pine seeds and caches of nuts gathered by squirrels, but at that time only Thoreau had synthesized these observations into a theory of the dynamics of white pines and oaks in the forested landscape of New England. Thoreau thereby established a framework for how to think about the interactions of white pine and other species with one another and with the physical environment, a framework akin to the modern concept of an ecosystem.

Thoreau noted that a few acorns in an oak forest nonetheless did survive the early frost and by the end of November were beginning to sprout. In his journal for November 1860, he wrote a memo to himself that it would be worthwhile to come back to these stands the next spring and see how many of these oak seedlings were still sound.[40] However, in the spring of 1861, Thoreau journeyed to the drier air of Minnesota in an attempt to cure a nagging respiratory affliction. Upon his return, his ailment developed into tuberculosis, from which he died on May 6,

1862, without having made any further notes in his journal about white pine, red squirrels, and oaks.

~

From Thoreau's three journeys through the Maine wilderness, there emerged a forceful rejection of his contemporaries' view that the virgin forests were there to harvest for the building of homes and a nation. In *The Maine Woods*, Thoreau put forth the very radical proposition that white pines, especially the large and old patriarchs, have a right to exist for their own sakes. Few people took him seriously at the time; further development of these thoughts had to wait until John Muir expanded upon them in his writings from the western pine forests of Yosemite and the Sierra Nevada. Today, almost all conservation, environmental, and wilderness organizations apply these ideas to not only white pines but all of Nature. But an important origin lay in Thoreau's distress over the absence of legendary white pine monarchs in the North Maine Woods.

In *The Maine Woods* and in his journal and later writings, Thoreau also glimpsed how white pine might be the biological foundation of the North Woods. He saw that the white pine's life cycle also includes dead and decaying snags and downed logs, which provide substrate for the mosses and seedlings that live upon them, and that red squirrels and oak seedlings depend on the mature white pine for their own existence. Thoreau's thoughts about the roles of giant trees like white pines in forest ecosystems would resonate strongly a century and a half later, when the loss of live, dead, and decaying old-growth Douglas fir forced ecologists to begin thinking about how spotted owls, mosses, and fungi in old-growth forests in the Pacific Northwest would be affected by a forest without trees.

Moreover, Thoreau's investigations of the dispersal of seeds of white pine, oaks, and many other species across the Concord landscape were the forerunner of what a century later would become landscape ecology, or the study of the causes and dynamics of spatial patterns of organisms across the land. Thoreau realized that the distances by which wind and red squirrels disperse pine seeds and acorns set the spatial scales for these dynamics and therefore the sizes of these patches and the distances between them.

Thoreau's writings were also the first deep investigations into the phenomenon of forest succession, which remains a core problem in American forest ecology. In the 1930s, the American ecologist Frederic Clements described succession as a linear sequence of species replacement leading eventually to a climax "superorganism" forest dominated by tree species that could reproduce in their own shade.[41] Clements took the term *superorganism* seriously and compared each species' role in the plant community to that of specialized organs in a body. In contrast to Clements's views, Thoreau thought of succession as not a one-way street to a stable climax but a dynamic cycle of pines and hardwoods replacing one another in patches across the landscape. But Thoreau's idea of cyclic patterns of succession in northern forests went unrecognized until ecologists Ross Wein and Mohamed El-Bayoumi independently rediscovered it more than a century later.[42]

~

As Laura Dassow Walls, Thoreau's most recent biographer, points out, Thoreau's moral and ethical ruminations about Nature and wilderness, like his scientific theories, were never independent of his natural history observations.[43] Thoreau's philosophical thoughts on the "higher law" of white pine in *The Maine Woods* and his natural history observations on the life cycle of white pine are excellent examples of how his moral philosophy of Nature melds with his natural history. For if a great pine has a right to exist on its own and not simply to be cut for lumber so that a team of oxen could stand on its stump, then the pine must be allowed to live out its life cycle, from the seed dispersed by the wind to the old-growth patriarch it will become.

Indeed, Thoreau pointed out that the life cycle of white pine extends beyond the end of its life to later stages, first as a standing snag, then as a dead and rotting log. All these stages of the dead pine provide structures and habitats for organisms, from mosses to woodpeckers, that also compose the North Woods. Thoreau would say that by executing our moral obligations to preserve the giant white pines, we also execute our moral obligations to the other organisms that biologically depend upon them. The higher law of Nature, the preservation of it, and the details of natural history of white pines and all other organisms were tightly interwoven.

As deep and prescient as Thoreau's observations of white pine's natural history were, they did not provide a quantitative understanding of how white pines grow and interact with other species. Such a quantitative understanding needed to be developed before practical attempts to restore white pine could begin. In the decades after Thoreau's death, George Perkins Marsh, Gifford Pinchot, Bernhard Fernow, and Volney Spalding would develop the quantitative descriptions of the ecology of white pine and the consequences of its loss that would serve as a foundation for a distinctly American forestry.

CHAPTER 5

The Watershed

If, instead of traveling north into Maine, Thoreau had gone first west into New York and then north up the Hudson River into the Adirondacks, he might have seen some of the last strongholds of virgin white pine in the eastern United States. The eastern slopes of the Adirondacks that drain into the Hudson River are steep and the peaks are high, extending above the tree line into one of the rare alpine ecosystems in the eastern United States.[1] These slopes of the Adirondacks were initially bypassed by loggers, who first harvested white pine from the gentler western slopes of the Adirondacks in the early decades of the 1800s.[2]

After draining the white pine, sugar maple, hemlock, and yellow birch forests of these eastern slopes, the Hudson River emerges from the Adirondacks near the small village of Greenfield. It was there, on July 10, 1823, just as logging was beginning in the Adirondacks and before Thoreau first went to Maine, that Sanford Gifford was born. Gifford became one of the major members of the Hudson River School of landscape painters, which included Thomas Cole, Frederick Church, and John Frederick Kensett. Of this group of artists, Sanford Gifford was the only one who lived much of his life in the Hudson River valley, first in the foothills of the Adirondacks, then in the town of Hudson in the Catskill Mountains downstream.

Gifford, like his fellow Hudson River School artists, was a precise, almost scientific, observer of nature. In 1871 he traveled through the Rocky Mountains as an artist with the Hayden Geological Survey, one of the expeditions that mapped the American West. But Gifford spent most summers in the North Woods of the Catskills, the Adirondacks, and northern New England to paint and record a landscape before and after it was logged, plowed, and farmed.

Most of Gifford's early paintings, as well as many of the paintings of the Hudson River School, portray the beauty and power of untouched Nature, especially in upstate New York and adjacent New England. The scales of most of these paintings, both in their physical dimensions as well as their scenic power, are majestic: they could easily serve as the backdrop to a Wagnerian opera. The North Woods of these paintings is what Americans like Gifford must have considered an idyllic wilderness, a North American Eden largely devoid of humans. White pines are often a prominent feature of these paintings, towering above all other trees. When people do appear in them, they are invariably tiny to emphasize the overwhelming power and endurance of the largely untouched mountains, rivers, and forests.

One particular painting of Gifford's, *Hunter Mountain, Twilight*, is different from most other Hudson River School paintings, however.[3] It tells a story of the landscape after the logging wave swept through. The logger's ax did not only cut the pines for the timber to make barns and homes but also cleared the way for the (white Americans') plows to follow. It is a story that would be repeated again and again across the northern tier of states straight through to the prairies. This is not a serene pastoral landscape of home and barn, dairy cattle grazing peacefully in green meadows, and corn shocks. Most of the painting is dominated by a small cutover bowl in the foreground. The stumps of trees in this bowl, most likely white pine and hemlock, are cut waist high, a common practice at the time but one that wasted valuable lumber.

There are no living white pines left. On the far side of the cutover bowl, a farmer herds his cows homeward to a barn, and the cows drink from a small creek, muddying it as they erode its banks and bed. The painting invites the viewer to consider the mud being transported downstream, clogging the gravel in which trout eggs and fry were developing. Gifford also invites the viewer to consider an even larger clear-cut immediately behind a fringe of remnant forest around the house and barn, ready for more plowing and further erosion. The landscape is in ruins.

In contrast to the wreckage of the forest and creek in the bowl and in the larger clear-cut beyond it, Hunter Mountain rises on the horizon in the distant and still uncut wilderness, a brooding blue mass beneath a luminous, lemon-yellow, twilight sky. Later, during the first two decades of the twentieth century, Hunter Mountain itself would be logged so heavily by the Fenwick Lumber Company that only a few patches of virgin forest would remain. Around the turn of the century, Winslow Homer would paint similar scenes of cutover forests in the Adirondacks as lumber operations worked their way up the Hudson and other rivers.

Some of Gifford's contemporaries thought that this painting portrayed the heroic pioneer carving a productive homestead out of the dark and dangerous wilderness. But others recognized the damage being done to the landscape by the eroding soil, the muddy stream, the eradication of the forest, the waste of good lumber left behind in the high stumps, and the loss of wilderness. *Hunter Mountain, Twilight*, embodies the period of time when Americans' ideas of Nature and their duty to it were becoming as radically transfigured as the landscape itself. Kirk Johnson, art critic for the *New York Times*, says *Hunter Mountain* and other paintings of the North Woods by the Hudson River School became a "force for conservation," both by idealizing the beauty and the power of the original wilderness and by illustrating the demise of it during settlement.[4] The ruined landscape left behind by the logging of white pine inspired many Americans to think critically about the connection between forests and the streams and rivers that drain them—the core idea of what we now describe as a watershed. Some of the first tangible efforts at the conservation of natural resources were attempts to repair these connections and restore the damaged watersheds.

George Perkins Marsh witnessed the destruction of watersheds much like that in *Hunter Mountain, Twilight* around his home in Woodstock, Vermont. A writer, diplomat, and linguist, Marsh was also a naturalist like his contemporary Thoreau. He saw that a watershed was not simply the river or brook that drains it or just the forests occupying its slopes but an intimate association between the two. With this insight, Marsh established the watershed as a primary unit of land conservation. In each denuded and eroded watershed, Marsh saw the erosion of civilization itself.

Marsh was one of those people whose broad erudition, clear thinking, and precise writing made others seek his advice on a wide variety of subjects. Marsh was at various times Vermont's railroad commissioner, State House commissioner, head of the Ethan Allen Monument Committee, and fish commissioner. It is in that last position that he began to think about how water moves across the landscape and how logging and farming affect it. Vermont is now, and was in Marsh's youth, well known for its spectacular brook trout streams and rivers. But by the 1850s, logging, construction of dams for lumber and grain mills, and the introduction of pickerel (a predator of brook trout) severely depressed the state's trout population. In an effort to reverse the damage to the streams and their fish communities, the governor of Vermont asked Marsh to make a report on how the brook trout could be restored. While artificial breeding and stocking of streams could certainly help restore the brook trout, the greatest threat to sustainable trout populations was, according to Marsh:

> the general physical changes produced by the clearing and cultivation of the soil [leave no doubt] that the spring and autumnal freshets are more violent, the volume of water in the dry season is less in all our water courses than it formerly was, and the summer temperature of the brooks [is] elevated. The clearing of the woods has been attended with the removal of many [natural] obstructions to the flow of water over the general surface [so] the waters which fall from the clouds in the shape of rain and snow find their way more quickly to the channels of the brooks, and the brooks themselves run with a swifter current in high water. Many brooks and rivulets, which once flowed with a clear, gentle, and equable stream through the year, are now . . . turbid with mud and swollen to the size of a river after heavy rains or sudden thaws. . . . In inundations, not only does the mechanical violence of the current destroy or sweep down fish and their eggs, and fill the water with mud and other impurities, but it continually changes the beds and banks of the streams. . . .
>
> The fish are therefore constantly disturbed and annoyed in the function of reproduction. . . . Besides this, the changes in the surface of our soil and the character of our waters involve great changes also in the nutriment which nature supplies to the fish, and while the food appropriate for one species may be greatly increased, that suited to another may be as much diminished. Forests and streams flowing through them, are inhabited by different insects . . . than open grounds and unshaded waters. The young of fish feed in an important measure on the larvæ of species which, like the musquito, pass one stage of their

> existence in the water, another on the land or in the air. The numbers of many such insects have diminished with the extent of the forests, while other tribes, which, like the grasshopper, are suited to the nourishment of full grown fish, have multiplied in proportion to the increase of cleared and cultivated ground.[5]

This is a remarkably complete and modern description of how water not only flows through a watershed but also connects aquatic and terrestrial food webs. The removal of first white pine, then spruce, followed by the destructive farming practices Gifford painted in *Hunter Mountain, Twilight* gave Marsh a large-scale thought experiment to probe how watersheds and their food webs work. Streams draining the uplands provide habitat for the larval stage of the life cycles of some insects, which then inhabit the surrounding forests during their adult stage. Cut down the forests, and not only does the hydrology of the streams change to the detriment of the spawning brook trout, but there is no more habitat to support the adult stages of mosquitoes, mayflies, caddisflies, stoneflies, dragonflies, and damselflies that would otherwise lay their eggs in the streams and support the brook trout. Instead, after logging and the conversion of the landscape to farming, grasshoppers dominate the upland pastures and croplands. Because grasshoppers have no aquatic life stage, they do little to support brook trout except for the few that happen to fall into the stream along its banks.

Marsh developed these ideas further in *Man and Nature*, which writer and environmentalist Wallace Stegner called "the rudest kick in the face that American initiative, optimism, and carelessness had yet received."[6] Marsh wrote this book in Turin at the foothills of the Alps while he was an envoy of the United States to Italy. In it, Marsh synthesized personal and scientific observations of deforestation in New England with his classical training in the history and languages of Mediterranean cultures, which had stripped the forests from the surrounding hills since ancient

Greece. *Man and Nature* is a massive clarion call for all nations to end the devastation of the land. In every one of *Man and Nature*'s more than 450 pages, you can hear the bugles calling Americans to action.

The term *watershed* and what exactly it meant was somewhat fuzzy in Marsh's time. Ever the linguist and classical scholar, Marsh explored the origins of this word in one of the many charming footnotes in *Man and Nature*.[7] Sir John Herschel, a noted astronomer and geographer, thought *watershed* came from the German *Wasserscheide*. Marsh disagreed. Instead, he proposed that the root word for *watershed* was the Anglo-Saxon *sceadan*, meaning both "to separate and divide" and also "to shade or shelter." A watershed separates and sheds the water that rains upon it, and the water then flows under the shade and shelter of the vegetation. In rooting the concept of watershed in the multiple poetic meanings of the Anglo-Saxon *sceadan*, Marsh recognized how complex this primary unit of the landscape a watershed is.

But Marsh went much further than simply providing a linguistic analysis of the origin of the word. If the First Law of landscape ecology is that water flows downhill,[8] then the Second Law must be: and it takes things with it. The lay of the land tells the water where to flow, and the flowing water strips bits and chunks of land away, shaping it further. Marsh makes the strong case that forests modify how water flows and thus moderate the discharge of streams. The key and revolutionary element of Marsh's scientific analysis of a watershed was a precise description of what happens to water flowing downslope, including the proportions made up by precipitation, drainage, absorption by vegetation and soil, and evaporation, and how humans can throw this flow of water through the watershed out of balance.[9]

After forests are removed, the land, bereft of its organic cover of vegetation and soil, no longer holds water in reserve but sheds it in flash floods in the spring. That is, as the absorption of water by trees, soil organic matter, and underground reservoirs diminishes, water moves quickly off the land, and the land progressively dries. Such drying proceeds slowly for generations, and that is its main danger, for it is not easily perceived until much damage has already been done. Eventually, the soil becomes so impoverished that it can be difficult to reestablish the forests that had heretofore nurtured the soil and protected the streams. It would be nearly a century before large-scale experiments began to test Marsh's framework of watershed analysis. Although later research further subdivided the pathways water takes, Marsh's analysis has remained the basic framework of forest hydrology ever since.

After presenting this analysis of watersheds, Marsh urged people not to wait until the deterioration was obvious but to begin to restore forests to the damaged lands right away. The forests, Marsh said, would not recover on their own but needed people to plant trees and manage them to restore the watersheds. At the same time, he urged preservation of tracts of lands whose forests had not yet been felled, particularly those at the headwaters of streams and rivers. He focused on New York's Adirondacks, specifically, to protect the headwaters of the Hudson River.

Marsh wrote that "the lofty woods [of the Adirondacks]. . . serve as a reservoir to supply with perennial waters the thousand rivers and rills that are fed by the rains and snows."[10] *Man and Nature* inspired New Yorkers to protect most of the remaining North Woods in the Adirondacks and Catskills where white pine had not yet been cut as "forever wild" forest preserves to deliver clean water to the Hudson River for the downstream residents of New York City.[11] This was perhaps the first time that a watershed was legally recognized as an integral piece of the landscape, a central feature that we needed to adjust our economy to, rather than imposing our economy and politics upon it.

~

Legal protections for the Adirondacks and Catskills would be just the beginning of a shift toward protecting and restoring white pine and managing it for more sustainable yield of both timber and clean water. Both Sanford Gifford and George Perkins Marsh had a remarkable effect on the life and career of Gifford Pinchot, the founder of the US Forest Service.[12] Indirectly, *Hunter Mountain, Twilight* and *Man and Nature* each played a role in Pinchot's establishment of the US Forest Service. The sketches for *Hunter Mountain* were begun in 1865, the year that Gifford Pinchot was born, and the finished painting was purchased the following year by James Pinchot, Gifford Pinchot's father. James Pinchot made his fortune by extracting timber from the North Woods but later in life realized a self-conversion to conservation and forestry. James named his son Gifford after the painter, and Gifford the painter became godfather to Gifford the son. *Hunter Mountain, Twilight* hung in prominent places in the homes of James and later Gifford Pinchot. The continuous presence of the painting in the family's homes must have reminded father and son what Americans were doing to the landscape; perhaps more important, it implicitly asked how we could change course.

Marsh's *Man and Nature* also made a lasting impression on Gifford Pinchot. For his twenty-first birthday, Gifford received a copy of *Man and Nature* from his younger brother Amos. In the introduction to his memoirs, *Breaking New Ground*, Pinchot calls *Man and Nature* "epoch making," especially since forestry did not "occupy any appreciable space in the American mind of Civil War times," when *Man and Nature* was published.[13] Until his death in 1946, Pinchot's goal was to make American forests and forestry occupy as much space in the American mind as it did in his own.

Shortly after Pinchot's graduation from Yale, his father bankrolled him on an educational trip to Germany and France to learn European techniques of forest management. On this trip, Pinchot realized that European methods of single-species plantation forestry would never do to repair the American forested landscape and manage forests sustainably for both profit and repair of watersheds. Americans wanted more from their forests than timber, including other products like maple syrup and habitat for game species as well as a place for recreation, which single-species plantations championed by German foresters could not supply. Instead, Pinchot realized that sustainable forestry and conservation in America needed to be based on knowledge of the natural history, biology, and ecology of American trees and forests.

Accordingly, Pinchot and his colleague Harry Graves, cofounder with Pinchot of the Yale Forestry School, the first forestry school in the United States, began to make a detailed study of white pine on the lands in Pennsylvania owned by the Phelps Dodge Corporation. This was the first formal study of the natural history and management of any American tree species. *The White Pine* is a short book; at only 102 pages long, it was intended as a handbook foresters could take into the woods with them. The first third of the book discusses in a general way the relationships among white pine and soils, fire, wind, and other tree species. The remaining pages consists of tables of volume, growth, and other measurements of white pine growing on different soils. In the white pine forests on Phelps Dodge land in Pennsylvania, Graves found one monarch white pine that was 351 years old, 42 inches in diameter, and 155 feet high, estimating that the merchantable timber from the trunk of this single tree would yield 3,335 board feet of lumber. Pinchot wrote, "I doubt whether the mate of this patriarch can be found alive today."[14] Presumably, the Phelps Dodge Corporation did not let this tree escape its sawmills, either.

Pinchot also dedicated much thought to the ecology of watersheds. His father had taught him to fly-fish in the brook trout streams draining the North Woods of the Catskills and Adirondacks, a skill and devotion he retained to the end of his life. In his essay "Time Is Like an Ever Rolling Stream," Pinchot wrote, "Men may come and men may go, but the Sawkill brook flows on—feeding its trout, protecting its insect, molluscan, and crustacean life—a home and a hiding place for myriads of living creatures—a thing of beauty and a joy forever. Along its banks, giant Pines and Hemlocks have germinated and grown, flourished and died, decayed and vanished, uncounted generations of them, each having its contribution to the richness and glory of the place."[15] This is eloquence worthy of Thoreau's *Walden* married to the watershed insights of George Perkins Marsh. Few American foresters since have written with the clarity and fluency of Gifford Pinchot.

~

Marsh made a passionate call for a better understanding of forests' relationship with water and through this the changes that occur in ecosystems as a whole if this relationship is disrupted. But Marsh believed that it would be very difficult, if not impossible, to make accurate measurements of each of the pathways that water takes as it flows through a watershed. Taking up Marsh's challenge to better understand how forests control the flow of water through a watershed, Gifford Pinchot put American scientists on the path to understanding these complex connections when he and Theodore Roosevelt established the US Forest Service.

After Pinchot had completed work as a private consultant for Phelps Dodge and other landowners, President William McKinley appointed him head of the Division of Forestry in the US Department of Agriculture. After McKinley's assassination, Theodore Roosevelt became president in 1901. Roosevelt had an abiding interest in wildlife and conservation, and he and Pinchot became close colleagues and friends.[16] The country's first national forests had previously been placed in the Department of Interior by Presidents Grover Cleveland and William McKinley. This put Pinchot in the odd position of being the nation's chief forester (a title he chose for himself with Roosevelt's concurrence) without oversight over the national forests. In 1905, Pinchot convinced Roosevelt to transfer the national forests to the Department of

Agriculture and to reorganize the Division of Forestry into the US Forest Service. Pinchot divided the Forest Service into two main branches, one that would manage the national forests for timber and other natural resources and another that would do the research necessary to provide a scientific basis for forest management.

In establishing the research branch, one of Pinchot's first actions was to create experiment stations to give American forestry a sound scientific foundation. The stations would illuminate the natural history and growth of trees, how they interact with their surrounding environment, and how different harvesting practices might enhance or harm the forest. One of these experiment stations, the Coweeta Hydrologic Laboratory, provided the first quantitative analysis of how water flows through watersheds and how the restoration of white pine in the uplands could control the flow of streams draining them.

Coweeta lies at the border of North Carolina and Georgia at the southeasternmost tip of a finger of white pine's range down along the higher elevations of the Appalachians. Logging of white pine as well as chestnut, oak, and tulip poplar on these steep mountain slopes in the years after the Civil War destabilized the soil and sent it rushing away in muddy creeks. After the Forest Service created the Coweeta Hydrologic Lab in 1934, researchers there began to investigate how reforestation with different tree species would stop this erosion. Accordingly, they built weirs (small concrete dams where they could measure water flow and take samples) at the mouths of streams draining the many small watersheds in these mountains. They also documented the species composition, population structure, and age of vegetation of each watershed to learn how streamflow and water chemistry differed in the various forests.

Farmers and loggers wanted to replant abandoned farmland surrounding Coweeta with white pine in order to restore a supply of valuable timber, but no one knew how it would affect streamflow compared with the more common deciduous hardwoods. Accordingly, in two of these watersheds, researchers cleared the deciduous hardwood vegetation and replanted them with white pine in 1956 and 1957. In both watersheds, streamflow declined steadily as the white pine canopy developed. Eventually, the streamflow was much less than from hardwood-dominated watersheds in the region. Streamflow was reduced more under the evergreen pines during winter and early spring compared with watersheds with a hardwood canopy.[17]

Why would the type of canopy affect streamflow? These experiments demonstrated that white pine had such a large effect on streamflow because its evergreen crown intercepted and held more snow and rainfall in fall, winter, and spring compared with the bare hardwood branches, which allowed snow and rain to fall to the ground; the water and snow subsequently evaporated directly off the pine needles.[18] Second, because it retains needles for at least two and sometimes three years, white pine has greater leaf surface than deciduous hardwoods and can release more water through the stomates, or pores, of its needles than can the crowns of oak or maple.[19] The findings at Coweeta have since been corroborated by other researchers.

The work done at Coweeta and at many other experimental stations that Pinchot created suggested that the loss of white pine from much of the North Woods may have greatly changed the streams flowing from them. Without the year-round canopy of green needles holding up rain and snow and releasing water brought up from the roots back into the atmosphere, the water must leave the watershed through the streams. It took nearly a century, but these watershed-scale experiments finally quantified the movement of water that Marsh had so carefully outlined. Such watershed studies are now widespread throughout the world.

In their paintings and writings, Gifford, Marsh, and Pinchot gathered up the loss of white pine and the subsequent erosion of the American landscape and poured forth the river of a new conservation movement. Until the mid-nineteenth century, white pine was simply a commodity, nothing more. As the environmental historian Bathsheba Demuth writes, "Convert too much of any species into a commodity and consumption exceeds reproduction."[20] What follows is a steady decline of the species being exploited. Gifford, Marsh, and Pinchot, as well as Thoreau before them, recognized that the commodification of white pine caused its demise and also made the natural landscape less productive, less useful, less beautiful, and a poorer habitat for other species.

Underlying the paintings and writings of these men was the nascent idea that an ecosystem is built from the intersections among the life cycles of its component species and the soil and water that nourish them. To restore damaged ecosystems like that in *Hunter Mountain*, Marsh, Pinchot, and others realized that we would have to develop a quantitative understanding of the life cycles of white pine and other species and then implement that new understanding with management plans. That new understanding would become the core of the sciences of ecology, forestry, and conservation.

CHAPTER 6

A Scientific Foundation of White Pine Ecology and Management

The dangerous felling and transport of white pine logs from the North Woods required a solid working knowledge of engineering, but it didn't require any scientific knowledge of white pine biology. Many of the harvested pines were near the end of their two-hundred-plus-year life spans. The loggers did not need to know how the trees grew to their large sizes, only where they were and how to fell and skid them to a river for the log drive to the nearest sawmill. The cutting and transport of logs and the fires that subsequently swept through the slash left behind a shattered landscape. Gifford Pinchot said repeatedly and in his usual direct manner that American forests had been "butchered" by the wholesale logging of white pine. One of the first jobs of the profession of forestry was to restore the North Woods to a more productive state. This meant restoring white pine back to the forest.

Pinchot and his acolytes wanted to restore white pine to the North Woods partly to reestablish a sustainable supply of lumber from the North Woods. To do this, Pinchot told his foresters in the fledgling Forest Service to think about a forest as something more than a vast warehouse of timber. To be sustainable, trees that are cut need to be replaced by the next generation, which will take many decades to mature to sizes that are worth cutting and transporting to the mill. To manage a forest sustainably, Pinchot's foresters needed to pay attention to the life cycle of trees, beginning with the seed and seedling stages on up through the sapling, cone-bearing, and mature stages and beyond, to "overmature" and then dead trees.

Other than Thoreau, previous naturalists and loggers ignored the little seedlings underfoot. If saplings were of any use to loggers, it was only to help cushion the fall of the giants towering above them. In today's forestry terminology, these early life stages are "pre-commercial"—that is, they don't make any money for loggers and sawmill owners. Nurturing these young trees incurs costs and risks that foresters hope to recoup much later, when the trees become large enough to be profitably harvested.

This dilemma still faces foresters today: the profits made from harvesting the valuable large and mature trees aren't realized during the decades when the costs and risks of establishing and nurturing young trees are incurred. These costs include investments in planting, pruning, and thinning. At the same time, the young trees face risks of loss or deformation by fire, insects, and diseases. A willingness to defer profits for decades while investing in the care of younger trees is what separates modern forestry from the era of the liquidation of the virgin pine forests.[1]

The dilemma of deferring profits while the trees completed the younger stages of their life cycles, not a desire to preserve and restore the wilderness, is what initially drove foresters to study the ecology of white pine. In the forestry profession, the study of the life cycles and the ecology of timber-producing species is known as silvics, and the application of silvics to timber management is known as silviculture. If silvics is the scientific foundation of forestry, then restoring white pine to cutover lands was the foundational problem of silvics in America.

~

Silvics was born in Michigan as the loggers were moving west into the great pineries of the Chippewa Valley in Wisconsin and farther into Minnesota. Probably no one did more to establish the scientific foundation for restoring and managing the white pine to the North Woods than Volney Morgan Spalding, professor of botany and curator of the Herbarium at the University of Michigan. Spalding was born in upstate New York in 1849. When he was still a boy, Spalding's family moved from upstate New York to a farm near Ann Arbor, Michigan. At that time, the epicenter of white pine harvesting had passed from New England and New York to Michigan. While growing up and later as a student at the new University of Michigan, Spalding watched the great wave of white pine harvesting that swept through his state, leaving the landscape and an important part of Michigan's economy decimated. These experiences made him wonder whether it would be possible to restore white pine and a thriving timber industry to Michigan.

This journey eventually brought Spalding to his professorship at the University of Michigan, deep in country built by the white pine logging boom. There, in the 1880s, he taught a course in forestry, the first such course taught anywhere in the United States.[2] The modern University of Michigan was founded in 1837 as part of the constitution of the newly created state of Michigan.[3] Unlike universities along the Eastern Seaboard, this university would not be structured according to the classics model of Oxford, Cambridge, and other British universities. Instead, it was to be the American leader of a Prussian model of public education, in which primary and secondary schools and universities were integrated into a seamless whole and supported by public funds. The curriculum emphasized research on how to solve practical problems facing the new state, whose economy at that time was supported largely by white pine logging.

This was the academic atmosphere in which Spalding thought about botany. Late in his career, on December 30, 1902, Spalding gave an address to the Society for Plant Morphology and Physiology, of which he was president. There, Spalding recalled that when he began teaching botany, he had only Asa Gray's *Manual of Botany* and a few other books.[4] As he remembered, "Microscopes, of a certain sort, there were, but no other apparatus whatever. Razors were sharpened on a well-hacked strap, iodine and sulphuric acid constituted the reagents, and the enthusiasm of fellow adventurers in an unknown country kept up the courage of young men and women who walked by faith and saw but little."[5] Outside of Harvard, where Asa Gray had established a world-class program in botanical research, botany at that time in the United States was regarded as being mainly "for edifying the minds of young ladies . . . and taught by instructors whose learning did not extend much further than *Claytonia virginica* [spring beauty] and *Erythronium americanum* [trout lily],"[6] two spring flora that were avidly sought, collected, pressed, and framed for hanging in drawing rooms. A rather primitive (and sexist) state of affairs.

As a professor of botany, Spalding sought to rectify this sorry condition and do what he could to shift the study of botany into a modern, quantitative science. True to the mission of the University of Michigan, Spalding emphasized that the university should justify the investment of public funds in basic and applied science by providing the next generation of students a broad training that would make them engaged citizens who would address societal problems. Although his presidential address was motivated by scientific issues, it also helped promote the idea of a liberal education for all citizens.

Spalding argued that nowhere in Michigan was this training of students more needed than in the application of the nascent field of scientific ecology to forestry, especially reforestation of the white pine land, "that has been the greatest source of wealth to the state [but] is now a wilderness, practically worthless until it is clothed again with forests." He was encouraged by the data even then being collected by professional foresters, some of whom had taken his class in forestry and were already working for Pinchot.

Accordingly, Spalding began writing a monograph on the life history, growth, and management of white pine that would describe and analyze its growth on different soils and in different regions, its interactions with other species, and how foresters could use this knowledge to restore and better manage white pine for a sustainable timber supply. However, Spalding's other administrative duties as department head, as well as his failing health, did not give him time to finish it.[7]

The task fell to Bernhard Fernow. Fernow had been the third director of the Division of Forestry in the US Department of Agriculture before Pinchot took over and renamed it the US Forest Service.[8] In contrast to Pinchot and Graves's small 1896 handbook on white pine management,[9] Spalding and Fernow's more comprehensive study, published three years later and titled *The White Pine*, was the founding scientific monograph of the ecology, silvics, and management of any tree species in North America. Unlike Pinchot and Graves's handbook, which was based mostly on white pine in Pennsylvania, Spalding and Fernow's monograph contained data and observations on the ecology and life cycle of white pine throughout its range.[10] The monograph was instrumental in describing the ecology of white pine and how foresters needed to consider it when restoring and managing white pine forests. It became the model for all subsequent studies of tree species by the Forest Service. These subsequent studies are now compiled in the Forest Service's two-volume *Silvics of North America*, also known as the Silvics Manual.[11] Every species entry in the Silvics Manual is organized along the lines of *The White Pine*.

~

Spalding and Fernow complemented each other well. *The White Pine* reflects Spalding's training as a botanist with interests in forestry and

Fernow's training as a forester with interests in botany and forest ecology. The subtitle of the book, *Pinus strobus* Linnaeus, is a clue to the reader that this monograph will have a strong botanical flavor. During the nearly 150 years between Linnaeus's species designation and Spalding and Fernow's book, virtually nothing had been studied about the natural history and ecology of white pine other than a few lines in Asa Gray's *Manual of Botany*, Pinchot and Graves's short book, and Thoreau's investigations, which hadn't yet been published.

This paucity of any previous research makes *The White Pine* all the more remarkable for its very modern and detailed description of the botany, ecology, and management of any tree species of North America. Unlike the discursive writing of Marsh and Darwin, and the sometimes-polemical writing of Thoreau and Pinchot, Spalding and Fernow's writing has the tone of a twentieth-century scientific monograph. It is hard to imagine Thoreau, Marsh, and Pinchot as contemporary colleagues, but I can imagine walking down the hall of my department to talk about forestry and forest ecology with Spalding and Fernow. As far as I am aware, there was no previous book like *The White Pine*. Although white pine is the focus of their book, Spalding and Fernow seem to be arguing something more general: in order to restore a species to a damaged ecosystem, we need to think systematically, quantitatively, and comprehensively about its own life cycle and where it intersects with the life cycles of other species. This has been the standard approach of ecology and conservation biology ever since.

The book begins with a sense of urgency, but also hope, articulating the importance of compiling what was then known about white pine and putting it in a form that could be used to manage and restore white pine to the North Woods. As was the practice at the time, Fernow sent a copy hot off the press to James Wilson, then secretary of the Department of Agriculture, the parent agency of the Forest Service. In a letter to Wilson accompanying the transmittal of this copy, Fernow writes:

> The situation regarding White Pine has materially changed since this monograph was first conceived, so that it might be charged that this publication comes too late. This would be a misconception. . . . The object of this monograph is to lay the basis for an intelligent recuperation of the virgin growth by the forest grower of the future, work which will surely be begun presently, but which would not be undertaken ten years ago. . . . This monograph is believed to be just in time for the use for which it is intended, namely, to prepare for the application of silviculture to the remnant of our pineries.

The White Pine is a blend of botanical and ecological observations with abundant tables and figures that provide detailed quantitative data on the growth of white pine in different habitats and at different layers in the forest canopy. From these observations and data, Spalding and Fernow offer practical advice for how to reestablish a white pine stand from seedlings; protect them from risks of injury from fire, wind, diseases, and insects as they grow to harvestable sizes; and maximize timber yield from them. Throughout the book, Spalding and Fernow relate the life cycle of white pine to various ecological factors such as climate, soils, and the light received by the canopy of an individual tree. This is now standard ecological theory taught in freshman college classes, but the reason it has become a standard approach in plant ecology is that Spalding and Fernow, as well as their contemporaries, pointed the way.

The book opens with a range map of the distribution of white pine in the eastern half of North America, showing not only the boundaries of its range but also its relative abundance within the range. This map is probably one of the earliest range maps of white pine and perhaps of any American tree species.

The creation of exact and quantitative range maps of species became one of the major tasks of naturalists, conservationists, and ecologists during the coming decades, and these maps became the basis of the science of biogeography. They helped ecologists ground theories of evolution and ecology in data on the distribution and abundance of species. Once ecologists knew the geographic distribution and abundance of a species throughout its range, they could then develop hypotheses about factors that control the growth and survival of its members. It is hard to believe that this most basic data about the ecology of white pine was not compiled until after two hundred years of uncontrolled logging had almost exterminated it.

Spalding and Fernow's detailed range map showed that the natural abundance, and by inference the growth, of white pine is not homogeneous throughout its range. Before logging decimated many white pine stands, it was most abundant from Maine and the Maritime Provinces westward through the Lakes States to Minnesota but tapered off north and south of this central zone with a long prong of lower abundance down the spine of the Appalachians.

Spalding and Fernow considered the reasons for the geographic distribution of white pine and its abundance to be one of the core problems of its ecology. In *The White Pine*, they suggest that the northern and southern borders of white pine's range are set by the climate, which determines the length of the growing season and the amount of heat or cold it is exposed to. At the northern border with the boreal forest, the growing season is too short and winters are too cold, whereas at the southern border with the southern hardwoods, the temperature may be too warm. Like the temperature of Goldilocks's oatmeal, the climate was "just right" in the middle of the range, where white pine was most abundant.

In chapter 1 we discussed the pollen record, which explains how white pine reached its current distribution as it migrated north and west while the climate warmed after the last ice age. The science of palynology, however, was only beginning when Spalding and Fernow wrote their book, and very little was then known about the geological history of white pine. Their ideas of how climate controls the current distribution and abundance of white pine across its range are remarkably prescient of much of the important research that would occupy palynologists and ecologists during the following century.

At a more regional scale, Spalding and Fernow showed that soil properties control the growth and distribution of white pine. White pine grows best and is most abundant on soils with fine sandy or loamy textures that can hold sufficient water to supply the trees. On coarser soils that hold less water, it is often partly replaced by red pine or jack pine, which can tolerate the increased intensity and frequency of drought. On clay and silty soils that hold more water, white pine is replaced by deciduous hardwoods, which are less drought tolerant but faster growing than white pine and so may outcompete it for nutrients and light.

At the finest scale, within a single stand of trees, the growth of white pine varies considerably from one tree to another. Spalding and Fernow claimed that the growth of individual trees within a stand depends primarily on the amount of light reaching its crown from above. Light is absorbed and dispersed as it flows downward through a forest canopy. Throughout its life cycle, a tree grows vertically up through this gradient of light: a seedling on the forest floor may be receiving only 5 to 10 percent of full sunlight or less, but as it grows upward, the amount of light bathing its needles increases until (should it survive) the needles at the very top of its canopy, one hundred feet or more above the ground, are in full sunlight. The tree grows into a better and better light environment throughout its life cycle.

This hierarchy of climate, soil, and light was a new way of organizing our thinking about the ecology of plants. It provided the foresters with an ecological framework within which to make decisions on how to manage pines throughout their life cycles within a stand, on different soils, and in different portions of its range. By harvesting competing trees around a white pine, the forester can readily manipulate the light gradient to speed its growth for later harvest, but how fast the white pine then grows depends on how well it can tolerate the climate and how much water and how many nutrients it can draw from the soil.

~

The risks to white pine are many. It is only by understanding how the life cycle of white pine meshes with the life cycles of fungal, disease, and animal species and natural disasters such as storms and fire that foresters can direct their efforts at reducing risk to the trees. Accordingly, Spalding and Fernow devoted ten of *The White Pine*'s eighty-one pages of text to these risks. The section titled "Dangers and Diseases" begins with brief discussions of wounds caused by "human agencies," including browsing by cattle (but not by deer, as deer populations were at that time much lower than they are today), storms, and fire, to which young white pines with thin bark were most susceptible. The recommended remedy for fire damage was to suppress it firmly and early. Ironically, this fire prevention policy is at odds with what ecologists later learned about the role of fire in white pine forests. (We'll discuss this further in chapter 9.)

Spalding and Fernow considered native diseases, mainly fungal species, to be greater risks to white pines. To control these diseases, they emphasized that foresters must understand how the life cycles of these fungal species intersect with that of white pine as well as alternative hosts and animals that disperse their spores. This laid the foundation for the modern understanding of the roles of diseases in ecosystems and food webs. Interestingly, no mention was made of the fungal species *Cronartium ribicola*, the white pine blister rust, perhaps the most widely known white pine disease today, probably because the rust was only entering North America when Spalding and Fernow's book was published. Nonetheless, as we shall see in the next chapter, to control blister rust, foresters and forest ecologists had to take Spalding and Fernow's approach to understanding its ecology.

In the next section of *The White Pine*, written by F. H. Chittenden of the Department of Agriculture's Division of Entomology, much more attention was given to insects associated with white pine. Again, the emphasis was on understanding the life cycles of insects in relation to that of white pine, including bark beetles and wood-boring beetles, the white pine weevil, and various leaf-eaters such as pine sawflies, moths, and plant lice. Chittenden did not consider any of these insects to be essential components of an ecosystem as ecologists would consider them to be today. Instead, he called them "enemies" of white pine—and he took the word *enemy* seriously indeed. For example, *Dendroctonus frontalis*, the most important of these insects, then and now, is discussed under the heading "The *Destructive* Pine Bark Beetle." (Emphasis mine, but Chittenden leaves no doubt where he stands on this species.)

The intricate weaving of the life cycle of an insect with its white pine host is nicely illustrated by Chittenden's account of the beautifully patterned butterfly the pine elfin. With a wingspan of only one inch, this related species is well named.

Although Chittenden designated this species as *Thecla niphon*, it is now reclassified into two closely related species, the eastern (*Callophrys niphon*) and the western (*C. eryphon*) pine elfins.[12] They are the only butterflies that use white pine to complete their life cycles. After hibernating in the crowns of white pines over the winter, the tiny adults emerge from their chrysalises in early spring and feast upon the nectar of flowers of blueberries, bearberries, and wild strawberries in nearby forest openings when not mating.[13] From April to June, males attract females by perching in full sunlight at the tops of pines adjacent to these openings. Upon mating, the females lay their eggs in the bases of this year's needles, one egg per needle. The caterpillars that later hatch from the eggs are as bright green as the new pine needles and

are thereby camouflaged from predatory warblers and other songbirds completing their spring migration to the North Woods. After feeding on pine needles, the caterpillars form chrysalises in late spring and wait for the following spring to metamorphose into the next generation of adults. Natural selection seems to have synchronized the emergence and mating of pine elfins with the spring flush of new pine needles and the blooming of spring flora. In this way, natural selection orchestrates the life cycles of these and other species into food webs.

To control pine elfin and all other insects, Chittenden recommended trimming and removing or burning infected plant parts, not realizing that natural fires played this role before the policy of fire suppression began. If burning could not be done, he recommended a liberal spraying of a soap solution consisting of one pound of Paris green (a copper-arsenic compound) mixed with a hundred gallons of fish oil, with a quart of carbolic acid thrown in for good measure. One wonders how many tons of arsenic from these sprays can be found across the range of white pine even today.

Unfortunately, *The White Pine* spoke of food web interactions only in terms of insect and disease "enemies" of white pine. Why or how white pine avoided being decimated by these enemies during the many generations prior to European settlement of North America does not seem to have occurred to these otherwise enlightened scientists. Until recent decades, many foresters followed Chittenden in having a negative view of other species in the food web if they decreased white pine growth and survival in any way.

~

If white pines survive all these risks, when should they be harvested? Prior to Pinchot and Graves as well as Spalding and Fernow, this would have been considered a silly question: the answer would have been "Whenever the logger can get to it." But when a modern forester has to nurture trees through younger life stages before they are large enough to harvest, the answer to this question becomes critical. Harvest trees when they are too young and you forgo additional growth (and lumber) the tree puts on in later years. Wait too long and you expose the trees to additional years of risk. And besides, old trees grow slowly.

To address this problem, Spalding and Fernow, and Pinchot and Graves before them, proposed that changes in the growth of a tree as it ages should determine when it should be harvested.[14] To analyze how the growth of the tree changes as it ages, Spalding and Fernow compared the current growth, called the current annual increment, with the growth averaged over the rest of its life to that point, called the mean (or average) annual increment. As a seedling grows into a sapling and then into a larger tree, its root system occupies a greater volume of soil to extract more water and nutrients, and its crown expands and captures more light for photosynthesis. Although the growth of a tree starts slowly, as it captures greater amounts of water, nutrients, and light with a larger root system and crown, its growth rate increases from one year to the next. Because of this slower growth at the beginning of a tree's life, mean annual increment lags behind the current annual increment. As long as the current annual increment exceeds the mean annual increment, it pays to let the tree grow and put on more and more wood faster and faster in each succeeding year.

But the current annual increment cannot increase forever. Eventually, as the tree gets older, its growth begins to slow as its roots occupy the bulk of the soil volume and as its crown begins to be constrained by the growth of the crowns of neighboring trees.[15] At this point, current annual increment peaks and begins to decline. For a while, it remains greater than the mean annual increment, so it still pays to let the tree continue to grow. Eventually, the current annual increment declines below the mean annual increment and begins to pull the mean annual increment down with it. The point at which the mean annual increment equals the current annual increment is called, rather mellifluously, the culmination of mean annual increment. After this point, the tree is producing wood more slowly than its average over its past, and it is time to harvest. This is known as the biological rotation age because it is based solely on the biologically determined patterns of tree growth and not the intended use of the harvested wood.

From data compiled from hundreds of trees grown on many different soils across the entire range of pine, Spalding and Fernow demonstrated that that the culmination of mean annual increment of white pine happens well into its second century but also depends on soil and climate. The poorer the growing conditions, the later the culmination of mean annual increment and the longer the rotation age. So long as the environment remains more or less the same into the future, harvesting at the biological rotation age ensures the maximum sustainable supply of

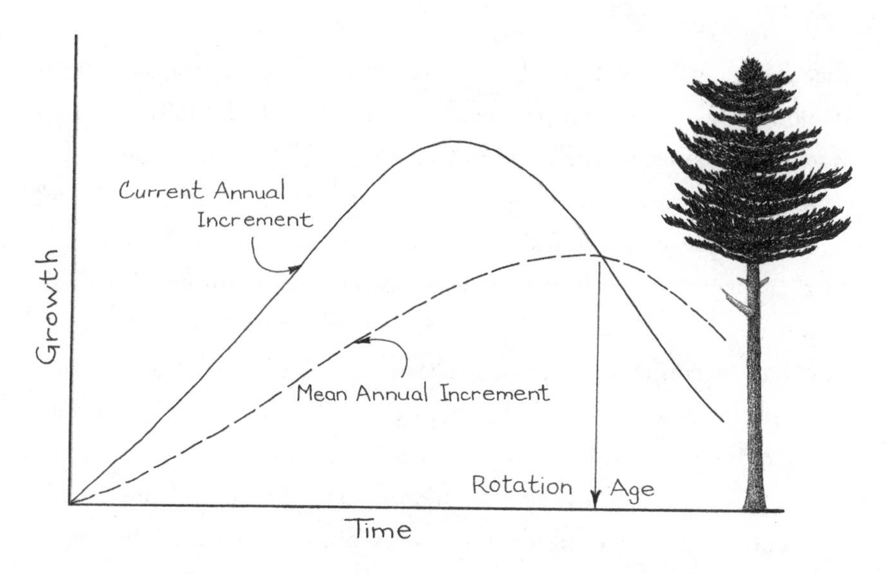

timber from that stand.[16] However, anything that diminishes the quality of a site, especially poor harvesting practices that erode or otherwise damage the soil, will extend the rotation age and expose the forest to increased risk of fire, storms, disease, and insects. *The White Pine* taught foresters that degrading and eroding the soil diminished and delayed profits by increasing both the rotation age and the risk of damage to the trees from insects, weather, and other agents at the same time that it destroyed the landscape. The destruction of the landscape was not simply an aesthetic problem, as we saw in the painting *Hunter Mountain, Twilight* by Sanford Gifford in the previous chapter. It was an economic problem, as well.

By accumulating data on the growth of white pine from throughout its range and especially on different soils, Spalding and Fernow gave foresters hard evidence of the impacts of soil health on their profits, forcing them to pay more attention to preventing erosion and soil degradation. Preventing erosion and the degrading the soil, and thereby protecting watersheds, was the major theme of George Perkins Marsh's *Man and Nature*. The genius of Spalding and Fernow's studies of white pine was that they showed practically how white pine, and a sustainable and profitable supply of white pine timber, could be restored while also restoring watersheds damaged by the erosive logging practices of the past.

More important, *The White Pine* pointed toward a more holistic and ecological approach to forest health than concentrating on harvesting a single valuable species and moving on as the supply was exhausted. To restore and manage white pine for timber and other benefits, foresters now had to consider the growth of white pine in relation to climate, soils, other tree species, insects, and diseases. Because Spalding and Fernow's monograph still saw white pine as a single focal species in the forest, it was not yet a fully ecosystem-based approach to forest management. Even now, more than a century on, we are just beginning to incorporate that approach, as we'll see in chapter 10. But *The White Pine* was the first step down that road.

Today, natural resource harvesting and management, particularly their profit-making aspects, almost always seem to be at odds with environmental protection. But it is useful to recall that conflicts between resource management and environmental protection were not always the norm. The fundamental goal of Pinchot, Graves, Spalding, and Fernow was to restore white pine and other species to the landscape in order to establish a profitable timber industry. These men held that research into the botany and life history of white pine, the geographic variation in its growth, and the biology of setting the rotation age were all necessary to put the timber industry in the North Woods on a sound footing that did not rely on a seemingly inexhaustible supply of huge white pines. Instead, Spalding and Fernow, along with Pinchot and Graves, wanted to restore forests that renewed themselves and provided clean water and habitat for game species while timber was simultaneously being harvested. Spalding, Fernow, Pinchot, and Graves integrated silvics and forest ecology with economics and policy. Ecology, sustainability, and profitability, in their view, were woven into a single tapestry that would restore a functioning forested landscape, yielding clean water and wildlife habitat as well as timber for local economies.

The White Pine laid the foundation for the development of silvics and forest ecology, providing a blueprint for scientific forest management for the next century. Other contemporaries of Spalding and Fernow, such as the mammalogist Joseph Grinnell and the limnologist Stephen Forbes, were also grappling with the problem of restoring other natural resources, such as game populations depleted by market hunting and lakes and ponds damaged by algae blooms.[17] Like Spalding and Fernow, these contemporaries compiled long-term observations and systematically gathered data on the lives of organisms in their native environment. All their research went far beyond observations by earlier ecologists such as Thoreau, expanding natural history to include the quantitative patterns of the growth of individuals and the intersection of the life cycles of organisms within food webs. The works of Spalding, Fernow, and their contemporaries have defined the methods and core concepts of ecology ever since.

CHAPTER 7

Rusty Pines and Gooseberries

After enormous fires that followed two centuries of logging, much of the North Woods was bereft of white pine. Many thought that white pine would never come back. The loggers had removed the largest and best-growing trees, which could have been seed sources for the next generation; whatever seedlings survived the logging were burned in the fires that swept across miles of land. Scientists and foresters were beginning to better understand the complexity of white pine ecology and soil health, but by the start of the twentieth century, no nurseries existed that could grow seedlings for reforestation, so the massive project of reforestation could not begin. Where, then, were foresters to obtain the seedlings they needed to restore white pine to its former range?

White pine seeds had been taken to Europe in 1605 by none other than Captain George Weymouth, whom we met in chapter 2 sailing off the coast of Maine. A century later, Thomas Thynne, First Viscount Weymouth and no relation to George, began planting white pine seedlings in abundance on his estate near Bath, England, hoping eventually to sell the large timbers to the Royal Navy when the seedlings grew to maturity.[1] Other estates planted white pine seedlings, and the handsome specimen trees that grew from them became popular for gardens of the landed gentry. French and German foresters also began introducing white pine into their forests for timber production. A sizable European nursery industry, producing hundreds of millions of seedlings each year, arose to supply these demands.

By 1900, American foresters and nurserymen were looking to these European nurseries to supply them with white pine seedlings. In the spring of 1909 alone, several million one- and two-year-old white pine seedlings and three-year-old transplants were imported from European nurseries.[2] These were used to start American nurseries and, in the case of three-year-old transplants, were planted directly into logged and burned areas.

Increasingly, however, as they opened crates of white pine seedlings imported from these European nurseries, American foresters began to notice odd yellow swellings on the branches of the tiny plants. These observations quickly came to the attention of Dr. Perley Spaulding, forest pathologist of the Bureau of Plant Industry of the US Department of Agriculture. After making several trips to northern New York and New England to inspect these seedlings, Spaulding identified these yellow swellings as one of the life stages of *Cronartium ribicola*, which was known in Europe but had never before been seen in America. *C. ribicola* is a fungal disease that infects all the five-needled white pines of the subgenus *Strobus*, of which eastern white pine is a member, but not the two- or three-needled yellow pines of the subgenus *Pinus*, which includes the jack and red pine that coexist with white pine. Because it affects white pine and its relatives, the disease quickly became known in America as white pine blister rust.

White pine blister rust is one of around seven thousand rust species known today, fungal diseases that infect plants, including many important crop and timber species. Spaulding traced many of the diseased seedlings back to one major nursery in Germany and then determined that seedlings from this nursery had been shipped in lots of five thousand each to 226 locations in New York, Pennsylvania, Vermont, New Hampshire, Massachusetts, Ohio, Indiana, Minnesota, and Ontario. Practically every five-thousand-seedling lot that Spaulding examined had some infected plants. In their well-meaning attempt to restore white pine, foresters and nurserymen also imported a disease that threatened to undo all their efforts and spread it throughout the North Woods.

~

It's hard to be a disease organism. You are always killing your own host and thereby cutting off the source of your sustenance. How, then, to survive the death of your host? Find another host to spend some of your life cycle in, preferably one that isn't harmed by you. After spending some time in your second host, you can disperse across the landscape and perhaps land onto individuals of the first host elsewhere.[3] Because of this struggle, diseases often couple the life cycles of at least two other species, each of which is necessary for the disease to complete its own life cycle. The complexity of almost every system of a disease and its hosts can be daunting. When blister rust was first introduced to North America, naturalists had yet to understand these intricacies.

Because blister rust always kills the pines it infects, it needs a second host to complete its life cycle.[4] This second host is any one of the species in the genus *Ribes*, flowering shrubs that include currants and gooseberries.[5] A number of species of *Ribes* in Europe already had blister rust, so when white pine seedlings were grown in Germany, the rust on European *Ribes* infected them.

There are about a hundred native species of *Ribes* in North America, including prickly gooseberry, Appalachian gooseberry, skunk currant (the fruits of which do smell like skunks), and American black currant. These native *Ribes* coexist with many of the other five-needled pines in the subgenus *Strobus*,[6] including western whitebark pine, western white pine, and sugar pine, the last two also being important timber species. So when blister rust was brought to North America from Germany, it could transfer to these other pine species through the native *Ribes* living around them.

While each of these *Ribes* species can host blister rust, they are rarely killed by it, because the rust infects only the deciduous leaves and does not spread into the twigs, stems, and roots of the plant. After the plant discards its leaves each fall (infected leaves seem to be shed even earlier), the rust no longer survives on them or in the forest floor—unlike most other fungi, rusts are not decomposers but require live hosts. The rust must therefore complete the *Ribes* phase of its life cycle within one growing season before the leaves are shed. During this phase, the rust goes through three stages of its life cycle. The first stage occurs in early spring when spores shed by the yellow blisterlike swellings on infected white pines enter the stomata of the young leaves of *Ribes*. Stomata are small pores on all leaves through which carbon dioxide enters and water and oxygen escape. They are closed when the plant is drought-stressed, so rust spores are most likely to enter stomata on *Ribes* where it grows wetlands or on heavy soils that can hold sufficient water so that its stomata remain open.

After a few weeks, orange-yellow spots or pustules appear on the undersides of the leaves. These pustules then produce different kinds of spores that spread to other leaves on the same plant as well as to neighboring *Ribes* plants. Several generations of this second stage, each lasting for several weeks during the summer, infect large swaths of *Ribes* but do not infect white pine. In mid- to late summer, the rust, which has by now spread to many neighboring *Ribes*, begins the third stage of this phase of its life cycle, producing brown, hairlike stalks on infected leaves. The tips of these stalks produce still different spores that cannot infect other *Ribes* plants but instead infect nearby white pines.[7]

White pines need to be nearby but not necessarily in the same stand to be within range of rust spores from *Ribes*. Although the prickly gooseberry is most abundant in uplands where it can coexist with white pine, other species such as skunk currant and American black currant are well adapted to wetter areas in lowlands where white pine typically is absent. Geographic distance doesn't always keep white pine from being infected, however. Convection cells sometimes form when the air cools on hilltops after late summer and early fall sunsets. Once the hilltop air cools, it flows off the uplands into the valley bottoms, where it can pick up rust spores from any infected *Ribes* growing there. When these air currents meet the wall of the forest or hillside on the other side of the valley bottom, they are deflected upward and spiral down the valley's axis, depositing the spores on the upland pines on either side of the valley.[8]

Once the rust spores land on white pine needles, they enter the needles' stomata. Since pine needles remain green and alive for two or more years before being shed, the rust can proceed through the pine phase of its life cycle at a more leisurely tempo than the race to complete its three stages in one season on *Ribes* leaves. The next spring, the spores produce mycelia, long rootlike strands, which enter the xylem and phloem vessels. These vessels are the tree's plumbing: xylem transports water and nutrients up from roots into the needles and phloem transports carbohydrates produced in the needles to the rest of the tree. These vessels are pathways for the rust's mycelia to spread throughout the tree.

Although it may take several years for the tree to die, the situation of the pine is now hopeless. After a year or two of spreading within the tree, blisters begin to appear on the outside of the bark. These blisters are the second stage of the pine portion of the blister rust life cycle and are what gives the disease its name. At first, the blisters are white, but they turn yellow when they erupt and release yellow spores. These particular spores can infect only *Ribes*, not other pines, and are released high in the crowns, where they are dispersed on the wind. Some fall back on *Ribes* within 250 miles from the original site of infection, and the blister rust life cycle repeats itself as the rust rapidly spreads across the landscape.

The different life cycles of its two hosts ensure the survival of blister rust so long as the spores produced on each host can get to the other. On the evergreen overstory of white pines, the mycelia of the rust develop slowly until blisters erupt through the bark, releasing spores that are widely dispersed by the wind flowing through the crowns. On the deciduous understory of *Ribes*, the rust races through its life cycle before the leaves are shed in the fall, and spores disperse a short distance to nearby white pines. Once the rust spores infect the white pine's evergreen needles, it persists through several winters until the pine dies. But because *Ribes* isn't killed by blister rust, the rust is able to survive the death of its pine hosts. This means that wherever *Ribes* and white pine live together across the same landscape, conditions are ideal for the long-term survival of blister rust.

The vast logging of the white pine combined with well-intentioned efforts to replant it created this ideal landscape. Most *Ribes* species grow best in areas where agriculture, fires, storms, floods, or other large disturbances open the tree canopy and allow full sunlight to reach the *Ribes* shrubs. The plants are spread further when species of songbirds disperse *Ribes* fruits and seeds into newly cleared areas. In colonial America, *Ribes* may have been locally abundant only where European settlers cleared the forests for farms and planted European black currant, which is particularly susceptible to blister rust. At that time, native *Ribes* appears to have been confined to valley bottoms and floodplains where periodic large floods killed or swept away trees and kept the forest canopy open. In his travels through the Northeast and Quebec in the colonial period, Peter Kalm, Linnaeus's student and botanical collector, mentions finding European black currant only in farm settlements near Philadelphia and Montreal; he does not mention wild currants or gooseberries at all.[9]

Even Henry David Thoreau, that ever-alert observer of nature, mentions *Ribes* only three times in his voluminous journals, and those only in passing.

However, the logging of white pine westward from New England during the 1800s and the fires that followed it created huge openings into which songbirds could spread gooseberry and currant seeds from these relatively few local sources. By the time American foresters began to plant infected European seedlings, the conditions for blister rust to thrive were already set. When infected seedlings were planted adjacent to native *Ribes* species or European black currants, the spread and persistence of the disease across the landscape was inevitable.

~

The white pine blister rust was the first widely recognized non-native species to become a pest in America with the potential to inflict major economic damage. The first step to controlling it was to stop the further importation of diseased seedlings of both white pines and *Ribes* from European nurseries. By 1912, only three years after Perley Spaulding identified blister rust on imported seedlings, Congress passed the Plant Quarantine Act.[10] Section 7 of this act specifically mentions white pine blister rust as one of its motivations.

The opening sentence of this act unequivocally stated that: "No person shall (1) import or enter into the United States any nursery stock or (2) accept delivery of any nursery stock moving from any foreign country into or through the United States unless the movement is made in accordance with such regulations as the Secretary of Agriculture may promulgate to prevent dissemination into the United States of plant pests, plant diseases, or insect pests."

This sentence is immediately followed by a listing of what those regulations will consist of, including the issuance of permits to move nursery stock into and through the United States, certificates of inspection, quarantine of imported stock to determine whether or not it is infected prior to further transport, and remedial measures the secretary of agriculture can take if stock is found to be infected. In addition,

> it shall be the duty of the Secretary of the Treasury promptly to notify the Secretary of Agriculture of the arrival of any nursery stock at port of entry, and no person shall import or offer for entry into the United States any nursery stock unless the case, box, package, crate, bale, or bundle thereof shall be plainly and correctly marked to show the general nature and quantity of the contents, the country and locality where the same was grown, the name and address of the shipper, owner, or person shipping or forwarding the same, and the name and address of the consignee.

Woe be to anyone who, with "case, box, package, crate, bale, or bundle thereof," would thwart the secretary of agriculture from performing duties owed to the American people by eradicating the blister rust from their forests. Quarantine Number 1, issued immediately on passage of the act, prohibited the importation of any five-needled pines; Quarantine Number 26, issued in 1917, authorized the eradication of any species of *Ribes* near white pine stands.[11] The Plant Quarantine Act, strengthened by further amendments, has remained in effect ever since and is the foundation of all further laws and regulations to control any invasive species.

~

The next step to restoring white pine without the danger of blister rust was to remove all *Ribes* plants near any white pines. To oversee this operation across the entire range of white pine, the Bureau of Plant Industry in the Department of Agriculture created an Office of Blister Rust

Control. Spaulding was put in charge of research on the blister rust's life cycle and how it was passed between white pines and *Ribes*. Initially, Spaulding thought that removal of *Ribes* within one hundred meters of white pines would be sufficient.[12] As field surveys began to demonstrate that this was not far enough to prevent spores from traveling from *Ribes* to pines, he progressively increased the *Ribes*-free zone to six hundred meters but then finally settled on a more practical two hundred to three hundred meters as a workable average.[13]

Samuel Detwiler, a plant pathologist also in the Bureau of Plant Industry, ran the operations phase of the program to eradicate blister rust. The geographic scope of this program was vast, extending from Maine west to Minnesota throughout the range of eastern white pine. Later, when blister rust began to infect the western white pine, the program was extended to the northern Rocky Mountains, as well. To implement this program over such a large area, Detwiler coordinated efforts by an organization of state and local foresters and plant pathologists called the Interstate Committee for Suppression of Pine Blister Rust.[14] Surveys for cultivated black currants were carried out by federal scientists from the Plant Quarantine Division of the Bureau of Plant Industry. These federal agents directed landowners to destroy their cultivated currant and gooseberry plots under penalty of law, but they sometimes met strong opposition.[15]

While the Bureau of Plant Industry originally supervised the field crews tasked with the eradication of wild *Ribes* species, the responsibility was later transferred to the US Forest Service. These agencies employed young men seasonally, either commuting daily from home to the woods or living in camps for the summer in remote areas in Maine, the Adirondacks, and northern Michigan, Wisconsin, and Minnesota. These young men experimented with various tools to dig up the *Ribes* plants and eventually settled on an army trench tool with the pick end split vertically in half to grab the stem at the root collar and leverage as much of the roots out as possible. This tool was dubbed a hodag after the mythical clawed beast in early Paul Bunyan stories. When the first group of young men went "over there" during World War I, they were replaced by teenaged boys.

During the Depression, the *Ribes* eradication program was re-created under the supervision of the Civilian Conservation Corps, which we'll discuss in more detail in chapter 8. The young men of the CCC probably removed approximately three hundred million *Ribes* plants from the nation's forests.[16] However, foresters also wondered whether white pine should be planted at all if it would only succumb to blister rust. Where blister rust was a lingering concern, the CCC chose to plant red pine, which as we learned earlier in this chapter is not susceptible to it. The red pines planted by the CCC men often survived better than white pine because they could also tolerate infertile soils. For these reasons, CCC plantations of red pine, almost all now approaching ninety years old, are common throughout the Lake Superior region. In fact, Jim Sanders and Dan Ryan, a retired supervisor of and a current ranger with the Superior National Forest in northeastern Minnesota, recently told me that they had not seen any white pine plantations throughout northern Minnesota until a few decades ago, when rust-resistant white pine seedlings became available.[17] The CCC men's preference for planting red pine in areas of blister rust probably expanded its abundance at the expense of white pine throughout the North Woods.

How effective were the eradication programs of the Bureau of Plant Industry and later the CCC in saving white pine? The evidence seems to be mixed. Eradication of the cultivated European black currant seems to have had the greatest benefit, partly because it was the host species most highly infected by blister rust and also because its control could be targeted to known farms with generally small plots of currants for home use or local sale.

Once the cultivated black currant was removed, further eradication of native wild *Ribes* species was not always effective, because all these species can resprout from roots left behind that the hodags did not catch. As few as ten to fifteen meters of *Ribes* stems per hectare can be a sufficient population for sustaining blister rust in a white pine stand.[18] Given that mature *Ribes* plants are multistemmed, with each stem approximately a meter long, this amounts to only a few mature plants per hectare. While the blister rust eradication programs may have been locally effective at protecting white pine, we will never know, because there seem to have been no large tests with proper pairing of treated and untreated control stands. Efforts at developing rust-resistant strains of white pine and spraying *Ribes* with herbicides also had mixed success with high economic and, in the case of herbicides, environmental costs.[19] Blister rust control programs were terminated by the late 1950s and early 1960s largely because the cost to maintain them was greater than the value of any white pine saved.

Nonetheless, the programs were instrumental in establishing the need for a sound scientific knowledge of the life cycle of white pine in relation to the life cycles of all other species with which it coexists.[20] At this point, the great national project to restore profitable and sustainable white pine forests had been churning since the turn of the century, when the Forest Service was established. The scientific study of blister rust, as well as Spalding and Fernow's monograph on white pine, marked a sharp turn in how Americans thought about managing species. As the developmental biologist and ecologist John Bonner puts it, a species is not simply a collection of similar individuals that interbreed; it is a life cycle, each stage of which interacts with particular stages of life cycles of other species.[21] The successful control of blister rust depended on detailed analyses of how its life cycle meshed with those of *Ribes* and white pine. The white pine–*Ribes*–blister rust system is an excellent example of how solutions to resource management problems both depend on and in turn spur basic research on the biology and ecology of organisms, food webs, and ecosystems. Once the details of the basic biology of these interacting species were known, then practical management solutions could be developed. Research on the white pine–*Ribes*–blister rust system and the management recommendations that followed became a model for all future research on forest diseases.

CHAPTER 8

Roosevelt's Tree Army

On April 17, 1933, John Ripley, a young recruit wearing an official olive drab uniform, climbed a pine in the George Washington National Forest in the Shenandoah Valley of Northern Virginia, cut the top off, and attached a rope and pulley to the trunk. Mr. Ripley then descended the trunk using the pine's branches like rungs of a ladder, cutting them off as he passed them. Upon reaching the ground, he and his fellow recruits attached the Stars and Stripes to the rope and hoisted it up this pine flagpole, thereby officially opening the first Civilian Conservation Corps camp, Camp Roosevelt, named after then president Franklin Roosevelt. And so began perhaps the largest forest restoration and conservation effort in the world's history and one of the most beloved government programs in the history of our country.[1]

~

By the 1930s, the vast acreages where white pine and other species had once stood tall were either cut over and burned or farmed and then abandoned. At the same time, the Great Depression had devastated the nation economically and socially. The genius of Franklin Delano Roosevelt was to develop a program that would simultaneously repair the poor conditions of the land as well as the economic and social structure of the nation. His solution was the Civilian Conservation Corps.

The large-scale program to eradicate blister rust begun under the Bureau of Plant Industry and continued by the CCC gave the government experience in organizing corps of young men to do the work required to help restore white pine across a broad geographic area. However, the basic idea of the CCC, like much of forest conservation in the early twentieth century, can be traced back to Gifford Pinchot. In 1911, shortly after President William Howard Taft fired him as chief of the Forest Service, Pinchot began an inspection of the condition of forests in the Adirondacks at the request of the Camp Fire Club of America. About half the 3.3 million acres inspected were owned by the state of New York, the rest by timber companies and other private associations and individuals. Pinchot was unsparing in his dismal assessment of the cutover lands.

His report came to the attention of New York's young state senator Franklin Delano Roosevelt. Roosevelt had an abiding interest in trees and forestry, employing a full-time forest manager on his estate at Hyde Park. Hyde Park contained many experimental plantations of different tree species, including a white pine plantation that Roosevelt planted in 1914 and which still existed as of 2005.[2] It also contained a virgin, old-growth, white pine–hemlock–mixed hardwood forest known as Kromelbooge Woods, where Roosevelt liked to escape.

In 1912, Senator Roosevelt invited Pinchot to speak to the New York legislature about the condition of New York's forests and those of the Adirondacks in particular. Pinchot went all out and gave a talk of which George Perkins Marsh would be proud, showing lantern slides of deforested areas in other countries and proclaiming that this could be the future of the American landscape. Pinchot's address galvanized

Roosevelt into becoming as strong a supporter of conservation and as much an acolyte of Pinchot as his cousin Teddy.[3]

Shortly thereafter, Pinchot began warning of an upcoming "timber famine."[4] After he was elected governor of Pennsylvania in 1922, he began thinking about the social implications of deforestation as well as the implications for timber supply, watersheds, and wildlife. Visiting many of the sawmill towns and hamlets left bereft by the decimation of white pine and hemlock, which had been abundant across the northern tier of Pennsylvania's counties, Pinchot wrote, "Discouragement and despair are written everywhere . . . in the faces of people as well as in the condition of the tumble-down houses and grass-covered streets."[5] As governor, he created a program to hire unemployed loggers, millworkers, and other rural poor to help plant trees and reforest the landscape. Pinchot's hope was that such a program would help both the land and rural communities recover. Human dignity and nature conservation would go hand in hand.

As governors of neighboring states, Roosevelt and Pinchot corresponded frequently. As Roosevelt's political career progressed, he began to think seriously about Pinchot's ideas of the benefits of reforestation to both society and nature. Within five days after his first presidential inauguration in March 1933, Roosevelt began framing a plan for what he christened the Civilian Conservation Corps, using Pinchot's Pennsylvania reforestation and social improvement program as a model. Initially, Roosevelt intended to employ mostly unemployed young men from the cities, thinking that getting these young men into the forests would improve not only the health of the forests but also the physical and social health of the men—and so it did. The men were fed three square meals a day in the CCC camps (better than they had at home in most cases), and despite the heavy physical labor, a typical CCC enrollee gained eleven pounds of muscle during the first three months of his six-month stint.[6] The men were paid thirty dollars per week, of which twenty-five was sent directly to their families back home. Altogether, twelve to fifteen million family members benefited directly from the $662,895,000 allotted to them from the enrollees' paychecks.[7]

Pinchot became an enthusiastic supporter of the CCC program. This in turn benefited his home state of Pennsylvania, which eventually had one of the highest numbers of CCC camps of any state outside of California. Pinchot himself visited many of the camps, with special attention to CCC Company S73, which replanted thousands of white pines, hemlocks, and other species in the Allegheny Mountains. Pinchot lobbied the White House to recruit rural as well as urban poor; as a result, the administration instituted a new policy that 55 percent of CCC enrollees were to come from small towns and rural areas if at all possible.[8]

The Department of Labor recruited and screened CCC enrollees, or CCCs as they became known, and then sent them to camps throughout the country administered by the US Army (with both reveille and taps) under the command of General Douglas MacArthur. The daily work of the CCCs was supervised by other federal agencies charged with land management. During the decade the CCC was in operation, three million men aged seventeen to twenty-three years were employed and housed in 4,500 CCC camps throughout the country.[9] This amounted to 10 to 15 percent of the men of this age in the country at the time.

The US Forest Service oversaw virtually all the forest management and reforestation work on federal, state, and even some private lands. More than half the CCCs spent their time doing timber management and reforestation of one sort or another. The CCCs planted 2.5 million acres of deforested and farmed-out land, mainly with conifers the timber industry desired but also with hardwoods where appropriate.[10] The seedlings were planted four feet apart along rows that were also four feet apart, which became known as a four-by-four spacing. At this density, approximately one thousand seedlings were planted in each acre, usually in one day, yielding an astounding 2.5 billion seedlings planted in ten years. To obtain the seedlings, the CCCs spent 6.1 million worker-days planting seeds in nurseries they established on federal and state lands, later harvesting seedlings from these nurseries to plant in logged-over or burned areas. In addition, the enrollees improved timber production and quality on four million acres by thinning, pruning, and selective harvesting. As the operational arm of the Forest Service's Smokey Bear policy of preventing all forest fires, the CCCs constructed over three thousand steel fire towers, some of which remain standing today, including a half dozen that currently watch over the pine forests of northern Minnesota.[11] Approximately 6.5 million worker-days were spent fighting forest fires, many of which were spotted from these fire towers. It is no wonder that the Civilian Conservation Corps was often called Roosevelt's Tree Army.[12]

These on-the-ground conservation activities slowed or even stopped erosion by reestablishing populations of trees growing on devastated land. The trees took up and stored carbon, repaired watersheds and thereby reduced floods, improved habitat for wildlife, helped renew a supply of building timber and pulp for paper, and provided many other long-term benefits to the American landscape and society. George Perkins Marsh's clarion calls for watershed and forest restoration were finally answered on a continent-wide scale, where it made a real difference.

In addition to the massive number of direct conservation improvements, the CCC program was also instrumental in educating a generation about the idea of conservation, something that Gifford Pinchot had long sought. Almost every CCC camp had a separate building or a portion of the mess hall that contained a small but well-chosen library of forestry and soil conservation manuals, natural history books and field guides, and wildlife ecology texts. In the evening, forest rangers and other "local experienced men" taught courses that enhanced and deepened the CCC's understanding of conservation they gained during the day, including basic forestry, wildlife ecology, plant and insect identification, and other topics. Fewer than 10 percent of the enrollees had any understanding or training in what conservation meant or how to do it when they entered the camps. Although these courses were voluntary (only a basic first aid course was required), in many camps over 80 to 90 percent of enrollees chose to take them. Perhaps even more important, forty thousand men who had been illiterate were taught to read in these evening classes. They were motivated to learn how to read by the abiding interest they were developing in forestry and natural resource conservation during their daywork.

These men became enthusiastic proponents of conservation for the rest of their lives. One enrollee wrote, "Our work is very interesting. Being out in the open most of the time, we learn more about nature and the natural resources we are trying to conserve." Another said, "The work we do in the Great North Woods gives us a greater understanding of what the word 'Conservation' really means. I am now a firm believer that conservation is necessary for the preservation of our forests." Some of the enrollees' comments in interviews approached the poetic: "[We worked] in a world alive with beauty more lovely than I had ever known."[13]

Many of these young men became founders of local chapters of conservation groups such as Ducks Unlimited and Trout Unlimited after they returned home from World War II. The habitat-improvement weekend activities of these chapters drew on what their members had learned during their time in the CCC. Planting trees in areas where white pine, hemlock, and other virgin forests had grown only a few decades before, improving the growth of trees by thinning existing stands, and seeing how these activities restored streams and wildlife habitat helped expand the political base of conservation beyond a few of Pinchot's acolytes.

As the CCC men began their massive project of replanting, their project was guided by the natural history observations and scientific knowledge of Pinchot, Spaulding, and many others. But foresters and forest ecologists were just beginning to learn another lesson about the interconnectedness of forest communities, this time below the soil. Without this knowledge, the seedlings the CCC men were planting would be threatened—and many wouldn't survive.

White pine's role as a foundation species starts when the pine seedling begins the first set of many interactions with the rest of the food web that will develop throughout the tree's life cycle. At first, a seedling is a tuft of needles attached by a slender, minute stem to roots penetrating the forest floor and soil. The needles begin the process of photosynthesis that will convert carbon dioxide into sugars to feed the roots, and the roots in turn take up the water and nutrients required by photosynthesis. Needles and roots can grow only to the extent that they receive needed resources from the other. This fine balance between needles and roots constrains their simultaneous growth and can even lead to a stalemate between the production of roots and the production of needles.

One way to break this stalemate is for the roots to increase their surface area to take up more nutrients and water so that the seedling can make more needles. To do this, all conifer seedlings form symbiotic associations with types of fungi called ectomycorrhizae. This symbiosis continues throughout the rest of the tree's life cycle. Many of the mushrooms we see and harvest (with great caution—some are very poisonous!) are the fruiting bodies of ectomycorrhizal fungi. The fungi attach themselves to roots and send out hair-fine, rootlike filaments called mycelia whose surface area exceeds the seedling's roots by many thousandfold. The fungi can often be seen as slight swellings on the root surfaces out of which the mycelia extend. Often, though, the mycelia can be seen only with a microscope. Supplied with carbohydrates from the needles, the mycelia take up nutrients and pass them along to the

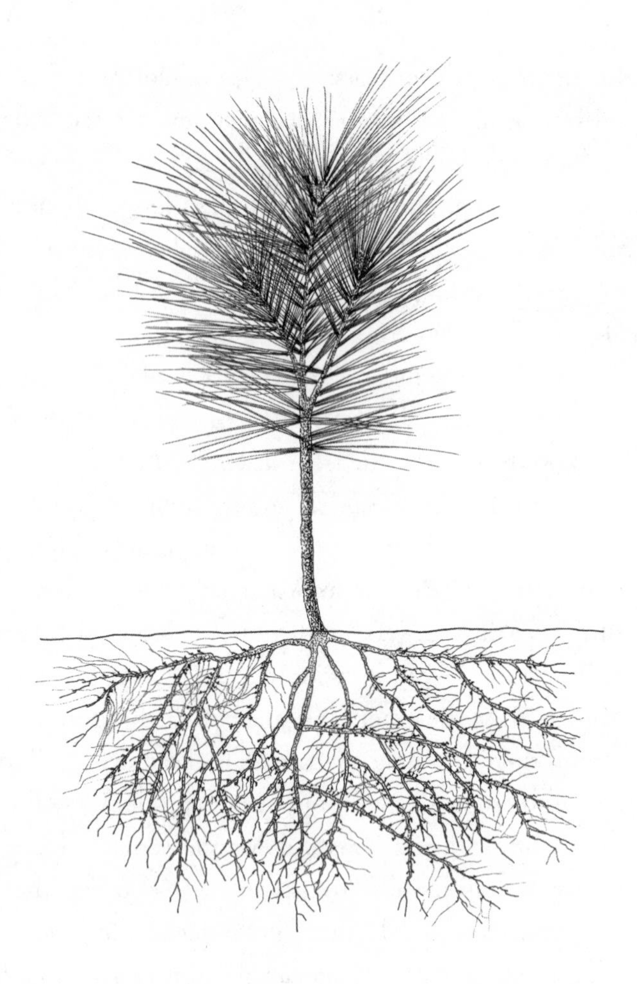

roots. Without the symbiosis, the white pine seedling often would die for lack of nutrients.

German botanists had discovered the connection between mycorrhizae and tree roots in the last half of the nineteenth century, but they weren't quite sure what this connection meant. One theory was that mycorrhizae, being fungi, were parasitic, which implied that they impaired tree growth. Another was that the mycorrhizae–seedling association was symbiotic and of mutual benefit. These two hypotheses were debated for several decades until experiments that confirmed the symbiosis hypothesis were completed in the 1930s by botanists H. L. Mitchell, R. L. Finn, and R. O. Rosendahl in the Black Rock Forest in Cornwall-on-Hudson in New York, using seedlings of white pine and white spruce.[14] Mitchell, confirming ideas of Spalding and Fernow (discussed in chapter 6), had previously found that the supply of solar radiation, the supply of soil nitrogen, and to a lesser extent the supplies of phosphorus, calcium, and potassium controlled pine growth.[15] Mitchell suspected that whether and how the seedlings obtained the soil nutrients depended on their degree of infection with mycorrhizae.

To study this, Mitchell, Finn, and Rosendahl grew pine and spruce seedlings in pots containing infertile soil with and without added nutrients and with and without mycorrhizae from soil taken from a nearby forest. Pine and spruce seedlings grown with mycorrhizal fungi formed mycorrhizal associations on over 90 percent of their roots, took up more nitrogen and phosphorus, and grew faster than seedlings grown without mycorrhizae. But seedlings grown in fertilized soil did not form mycorrhizal associations even when mycorrhizae were added. This surprising finding suggested that somehow the seedlings in fertilized soils did not need to form symbioses with these fungi, perhaps because they could obtain sufficient nutrients on their own. Several decades later, in 1959, Edward Hacskaylo and Albert Snow of the US Forest Service discovered that white pine seedlings formed mycorrhizal associations only when grown on infertile soils in full sunlight but not in shade. Apparently, white pine seedlings form symbioses with mycorrhizae only when the soil is infertile and when photosynthesis in full sunlight is high enough to provide both the seedling and the fungi with carbohydrates.[16]

Unfortunately, mycorrhizae were probably absent from the soils of many of the nurseries and pine plantations that the CCC planted. The nurseries were often established on abandoned farmland. Mycorrhizae that form symbioses with conifers do not do so with crops, and without extra carbohydrates from their preferred tree host species, the symbiotic fungi die when the soil is farmed for several years. Similarly, when nurseries and plantations were established in areas where fires burned through logging slash, they also failed, because the fires were often hot enough to burn off the topsoil and kill mycorrhizae. The seedbed, therefore, was not always hospitable to the pine seedlings the CCCs planted, and some of these nurseries and plantations were doomed to fail for lack of sufficient mycorrhizae in the soil. But I could find no reports of the CCCs returning to areas they planted to determine the survival of seedlings.

Unbeknownst to the CCC, the reestablishment of white pine and other species had to proceed hand in hand with the reconstruction of the soil food web, including the presence of a diverse collection of mycorrhizal species. Scientists wouldn't fully understand how mycorrhizae spread across the landscape for several decades after the CCCs planted their seedlings. In some cases, this dispersal happened by sheer luck, facilitated by natural processes of the forest food webs—but sometimes it did not. Many mycorrhizae, including those that most commonly form associations with pines, are spread through an intricate association with small mammals. While some mycorrhizae produce fruiting bodies above the soil—familiar forest mushrooms—and their spores are dispersed by the wind, many of those that associate with white pine produce fruiting bodies belowground. These fruiting bodies are eaten by many small mammals; in the North Woods, red-backed voles are especially important consumers of mycorrhizae and dispersers of their spores.[17] The spores are not digested by the voles, but remain viable in their fecal pellets, which they disperse throughout the forest. Once dispersed, the spores can come into contact with the roots of conifers such as white pine or white spruce, and the conifer–mycorrhizae symbioses form.[18]

However, the voles themselves depend on large downed logs on the forest floor for shelter and cover against owls and other predators overhead. This sets up a three-way interaction of mutual benefit to pines, voles, and mycorrhizae: the dead logs from previous generations of pines provide shelter and cover for the voles, the voles disperse spores from belowground fruiting bodies of mycorrhizae, and the mycorrhizae form symbioses with the next generation of pine seedlings. Because downed logs were lacking in the abandoned farmlands and burnt-over logged areas where the CCC planted pine seedlings, the habitat was poor for the voles that would disperse mycorrhizae into these soils, further threatening the survival of the planted seedlings. Without the seedlings to provide nutrients for mycorrhizae and without downed white pine logs to shelter the voles that disperse the seedlings, the North Woods could not recover in many places where logging, fires, and farming damaged the soil. And without the mycorrhizae, the white pine seedlings would not grow.

Recent research on mycorrhizal associations with white pine has also challenged the standard view that food webs are composed of chains of plants, herbivores, and predators. At least one species of mycorrhizae lures springtails, kills them with a toxin, and then decomposes them with its enzymes, slurping up nutrients from the decaying springtail carcasses. Up to 25 percent of nutrient uptake by the white pine seedling can come from springtails killed by this one mycorrhizal species.[19] The white pine–mycorrhizal symbiosis is, in at least this case, more like that of carnivore–prey rather than that of plant–soil.

By forming symbioses with mycorrhizal fungi, which then prey upon insects in the soil food web and whose spores are dispersed by voles that require white pine for habitat, white pine is therefore not simply a single, isolated species. These complex interactions weave together food webs above and below the ground, illustrating some of the ways that white pine is a foundation species underpinning the recovery and sustainability of the whole North Woods. The men of the CCC couldn't have understood how the seedlings they planted were integral to this complex web, and even modern research on mycorrhizae and their host trees continues to revolutionize our understanding of ecosystems.[20]

The CCC effectively ended as America entered World War II. In fact, because many of the CCCs were in superb physical condition and had learned to work together in a quasi-military structure, they became the core of recruits into the armed forces. Roosevelt tried to make the CCC permanent because there was much more conservation work to be done after the troops came home, but Congress would not agree. Besides, the country entered a new era of prosperity after 1945 that undercut one of the reasons the CCC was created in the first place: to help unemployed young men find purpose and meaning in a life of service to the country and to nature. The cessation of the CCC after World War II had implications for how the pines they planted continued to grow.

The seedlings the CCCs planted in the late 1930s and early 1940s—those which coincidentally grew in mycorrhizae-rich soil and were able to survive—would have begun growing into saplings in the 1950s. Once again, the life cycle of the white pine would have an important impact on how quickly these saplings could grow into a profitable harvest.

As a white pine grows from the seedling stage to the sapling stage of its life cycle, growth is increasingly allocated to the woody branches and the bole, or trunk.[21] The allocation of growth to the branches and bole has a cost that can be best understood by considering the carbon balance of a tree. Carbon is approximately 50 percent of the dry weight of any plant, more than any other nutrient. A pine's growth is therefore a balance between the net intake of carbon dioxide by photosynthesis in the needles; its conversion to carbohydrates that are then transported to the roots, bole, and branches; and the simultaneous respiring of carbon dioxide back to the atmosphere from these three organs. Roots, of course, obtain the water and nutrients needed for photosynthesis, thereby repaying their respiratory cost to the plant. Similarly, the bole and branches lift the needles higher into the canopy, where there is more light for photosynthesis. So while there is a respiratory cost to allocating growth to the bole and branches, the ability to raise needles to full sunlight repays this investment. When photosynthesis is greater than respiration, the tree is taking in more carbon to produce sugars than it releases and carbon is then stored in the tree, mainly in the bole, and it grows. To achieve the greatest growth of the bole, which they will harvest, foresters need to maximize photosynthesis by needles and minimize respiration by the bole and branches without sacrificing production of merchantable wood. This delicate balancing act is the main goal of silviculture.

To achieve this in the CCC plantations, it was expected that some future CCC crews would return to the plantations twenty years later to prune and thin the trees once they became saplings. Pruning aids the growth of the tree because the lower branches of the crown are shaded by the higher branches and so are using nutrients but not photosynthesizing effectively. In addition, if lower branches are left on the tree, the wood grows around them and they become knots. Pruning of lower branches happens naturally, but pruning by hand removes these lower branches faster and helps the tree produce lumber clear of knots earlier in its life cycle. Using extremely sharp polesaws, loggers and woods workers cut branches from the lower sixteen feet of the bole (a neck- and backbreaking ordeal). The resulting trunk will become a sawlog when later harvested, with clear, straight-grained lumber that can be sold at a higher price than knotty lumber. In addition to pruning, these plantations should also have been thinned once they reached the sapling stage to reduce competition from neighbors and enable the trees judged most likely to produce straight-grained and clear lumber to thrive.

Trees removed by thinning and the pruned branches have almost no commercial value. Pruning and thinning are costs borne by the landowner early in the life cycle of the tree in the hopes that he or she will be repaid with high-quality lumber when the mature tree is harvested many decades later, if in fact it survives. As we learned in previous chapters, the costs of forest management, such as pruning and thinning, are necessarily borne up front, but unlike almost any other investment, the profits are deferred for many decades, making timber management one of the riskiest ways to invest money. This peculiar economic risk has its roots in the long life of the tree and in the changes in growth during the transition from seedling to sapling.

When the CCC men came home from World War II, returning to the woods was not their top priority. The country's high prosperity was sparked in part by GI Bill grants to some of these men to go to college or to make down payments on houses, making woods work less attractive. It should be noted, however, that these benefits were available only to white men and were denied to African American and Native American workers and soldiers. But with the end of the CCC, opportunities for woods work were closed to all former CCC men, regardless of race or ethnicity.[22] In the meantime, the white and red pines the CCC men planted continued to grow. Unfortunately, without thinning, their growth also stagnated as competition between neighboring trees intensified—a process that continues today. The net result is that many CCC pine plantations are now dense stands of tall, slim trees of the same size and height that are not dying but not growing well either.

The migration of many of the CCC men and their young families from farms and cities to the suburbs sparked a great demand for lumber for new houses. The CCC red and white pine plantations were not yet ready to harvest and would not be for many decades. In the interim, to meet the demand for lumber from carpenters like my father, white pine was harvested from older second-growth stands that became naturally established from seeds dispersed by wind and squirrels from trees that had not been cut during the logging era. These pines had not yet self-pruned their lower branches, and had not been pruned by humans, and so lumber sawn from them was knotty and of lower quality than the clear lumber that came from the original virgin white pine forests, in which the old trees had long since self-pruned their lower branches. As the growth of the suburbs exploded, white and red pine supply could not keep up with the demand for new houses. Instead, the demand for lumber became increasingly satisfied by the yellow pines of the southeastern United States, particularly loblolly pine. Carpenters like my father did not like the "new" yellow pine lumber because the higher resin content made the wood harder to hammer nails into it and also gummed up their saws.

Despite these qualities, yellow pine lumber became more common in the lumberyards. Loblolly pine in the warmer southern states grows almost twice as fast as northern white pine and can be harvested much more quickly—twenty-five to thirty years compared with greater than seventy years for white pine. This means that, for loblolly pines, the costs of pruning and thinning pay off economically more quickly, without the additional risks of growing trees for many more decades. Plantation forestry is therefore the standard way to grow loblolly pine in the South, but no longer in the North. New plantations of white and red pine in northern states would never again approach the numbers established in ten years by the CCC.

~

The long-term results of the CCC planting campaign underscore a crucial point: while the CCC restored white pine and red pine to cutover lands in the North Woods, it did not restore pine forests. A plantation is not a natural forest and it never will be; it is a monoculture of evenly spaced trees instead of a naturally diverse mix of tree species providing habitat for many other species. Of course, the reforestation efforts by the CCC were never intended to restore pine forests; their purpose was to restore a supply of pine timber and restore the lives of unemployed young men. That the CCC succeeded spectacularly in enriching the lives of many young men is unquestioned.

It remains to be seen whether the promise of renewed timber supply from CCC plantations can still be fulfilled, but if they can be actively managed to reverse their stagnation, then there is some cause for optimism. In recent years, foresters have begun to thin CCC plantations in national and state forests in the hope that clear sawlogs can be harvested from them in a few decades. Prescribed burnings are also sometimes implemented to rid the understory of accumulated fuel and prevent devastating crown fires. As a result, more than fifty years after this process should have begun, the CCC plantations are slowly being nudged toward more economically valuable conditions.

As time goes on, however, it becomes more apparent that the CCC plantations would never grow to resemble natural pine forests even with more intensive management. If the vast plantings of pines by the CCC could not restore northern pine forests, even in places with mycorrhizae, then how did the original forests get their start in the first place? Before and during the great pine logging era, no one thought to ask this question. But in the years since the CCC seedlings were planted, many scientists and conservationists wondered what additional elements the massive campaign was missing. One answer to this question would be found by a young forest ecologist returning home from World War II as he studied the few remaining large tracts of virgin pine forest in northern Minnesota.

CHAPTER 9

Rebirth by Fire

While growing up in the 1950s and 1960s, I was inspired by Smokey Bear. In his most famous poster, issued in 1947, Smokey wore a campaign hat and jeans, held a shovel in his left hand, and, looking directly at us, pointed his right index finger straight at our faces and said, "Only you can prevent forest fires." Even more than preventing fires, Smokey seemed to call me to a life of work in the woods. I wonder how many ecologists of my generation felt similarly called. Smokey's pose and his message were copied directly from the US Army's recruiting poster of Uncle Sam, which told young men of the World War I generation that he wanted "you" for the US Army. Smokey told us that we were at war with forest fires and each of us had to do our part. This message resonated strongly with members of my parents' generation, and they made sure we knew it too.

There was good reason for this message. The CCCs of my parents' generation fought fires and built fire towers. The parents of my parents' generation knew, and in some cases experienced, the enormous fires that burned through white pine slash in Hinckley, Baudette, and Cloquet, Minnesota, and *their* parents knew and experienced the Peshtigo and

Thumb Fires in Wisconsin and Michigan. Huge fires in the white pine woods (or more precisely, in the logging slash left behind) were embedded in the memories of these three generations. Fires meant destruction of forests, homes, and lives. People who fought them and staffed fire lookout towers were heroes.

By many accounts, Smokey was the most successful public advertising campaign in history.[1] Forest fires decreased from thirty million acres per year before the advertising campaign began to less than five million in the years immediately after.[2] However, while Smokey's message greatly helped prevent fires started by accident or carelessness, its power and emotional appeal had an unintended side effect: it kept the Forest Service and many scientists from investigating the ecology of natural wildfires for quite some time. As we saw in the previous three chapters, foresters and forest ecologists had learned much about white pine in the seventy years or so since the great era of white pine logging ended. But their successes made them think that we could obtain a "perfect" forest that would thrive forever simply by planting, management, and protection. This idea ignored the important roles of natural disturbances, most especially of fire, in the natural forest. By protecting all forests from all fires, the Forest Service inadvertently set up an unnatural situation that continues to plague us.

In contrast to the fire suppression policy of the Forest Service, members of Anishinaabeg tribes in the region had long recognized that fire is an important and sometimes useful tool. They noticed that blueberries often grew in burned areas and sometimes intentionally set fires to prevent pines from invading blueberry patches, thereby maintaining blueberry plants that would sprout again from their roots.[3] Although the Anishinaabeg peoples and other Native Americans living in the North Woods before European settlement increased fire frequency above background levels set by lightning, those fires were largely small, local, and near their settlements and villages.[4] To American ecologists of the 1950s and 1960s, for whom fire was a destructive force, these ideas would have been hard to believe—but that was about to change.

Ironically, it was an ecologist with the Forest Service, Miron Lee Heinselman, who changed the way ecologists think about forest fires. Known to his colleagues and friends as Bud, Heinselman wrote relatively few papers during his career by today's standards, but each one had a monumental effect on the field. I knew and admired Bud; the way he thought about the North Woods greatly influenced my own thinking during my career as well as that of many of my colleagues.

Bud came from long lines of German and English farmers who stitched themselves together by marriage as they farmed their way westward from Pennsylvania and Virginia to Iowa. Bud's father ended this westward migration of farmers when he moved north to Duluth, Minnesota, began work as an insurance agent, and married Helen Krueger in 1917. Bud was born three years later, in 1920. Numerous family camping trips and "countless days roaming the wild range of cliffs overlooking Lake Superior near Duluth" fostered a love of the North Woods that would remain with him for the rest of his life and career.[5]

Heinselman's love of the North Woods eventually led to a PhD in Forestry in 1961. His PhD thesis was a study of black spruce growth and the hydrology of the enormous peatland around Red Lake, a remnant of glacial Lake Agassiz in northwestern Minnesota. This peatland comprises different plant communities arranged in curiously regular patterns across vast swaths of the landscape. Heinselman's thesis and the two papers based on it explained that these patterns are caused by the different ways plant species both respond to and control water flow.[6] Few ecologists had studied landscapes on such a large scale, but Heinselman's research on these peatlands stimulated ecologists to consider processes that might control whole landscapes rather than single plant communities. Heinselman then applied this way of thinking to understanding the history and role of fire in the pine-dominated forests along the Minnesota–Ontario border, known today as the Boundary Waters Canoe Area Wilderness.

The Wilderness Act of 1964 provided the impetus for Heinselman's research on the fire history of pine forests in the Boundary Waters.[7] The rivers and lakes of this area, which now trace the Minnesota–Ontario border, formed the main canoe route followed by the voyageurs from Lake Superior into Manitoba and points farther north and west. The Boundary Waters contains more water than land, distributed among more than one thousand lakes and ponds. In many places, it seems as though there is only enough land to keep the lakes separate from one another, making it easier to canoe through the Boundary Waters than to hike through it. The abundance of lakes and the lack of any major rivers to transport logs prevented the wholesale cutting of white pine well past when the rest of Minnesota's forests had been logged at least once. Much of the one million acres of the eventual wilderness area contained abundant uncut white and red pine forests, along with forests dominated by other conifers such as white spruce and balsam fir, with minor amounts of sugar maple, aspen, and birch sprinkled here and there.

Because of the spectacular beauty of the lakes and islands; an abundance of moose, wolves, beaver, loons, and other charismatic animals; the magnificent old-growth pine forests; and the lack of roads, the Boundary Waters has been recognized as a special landscape worthy of protection ever since the Superior National Forest was created in 1909.[8] In 1926, the general area of what is now protected wilderness was named the Superior Roadless Area. On the Ontario side of the international boundary, a matching wilderness landscape was then known as the Quetico land. Quetico–Superior was and still is the term used for the entire region.[9] In 1934, President Roosevelt created the Quetico–Superior Committee to advise the Forest Service on management of this area and whether logging should be prohibited in some places. But the difficulty of getting into and removing timber prevented many, although not all, of the old-growth virgin forests from being logged.

The 1964 Wilderness Act famously defined wilderness as "an area where the earth and its community of life are untrammeled by man, where man himself is a visitor who does not remain." Because of past logging and motorized boat access in some of the Boundary Waters, section 4(d)(5) of the act allowed these activities to continue there, but these exceptions were vociferously disputed by many conservation organizations.[10] In order to resolve the issues of how the Boundary Waters should be managed, Orville Freeman, then secretary of agriculture, appointed a committee under the chair of George Selke, former director of the Department of Conservation of the State of Minnesota. In its report to Secretary Freeman, the Selke Committee recommended that "the objective should be, in the main, to obtain a forest of the long-lived species, such as the red pine, the white pine, and the white spruce."[11] With this statement, white pine became one of three tree species that defined the wilderness character of this region. The problem was that virtually nothing was then known about the ecology of white pine or other forests in the Boundary Waters Canoe Area.

Enter Bud Heinselman. After he received his PhD, Heinselman became a research scientist with the Forest Service's North Central Forest Experiment Station in Saint Paul and Grand Rapids, Minnesota. In response to the Selke Report, Heinselman and several colleagues wrote a plan for a research program that would provide a scientific foundation for the management of the Boundary Waters as wilderness.[12] The plan focused strongly on the role of fire in forest dynamics, which Heinselman made the focus of his work for the next decade or so of his career. Through his studies, he illuminated the integral role that fire played in sustaining rather than destroying this forested landscape.

How did Heinselman know to focus on forest fires in white pine and associated forests of this region? As we shall see, his work was groundbreaking, but he was not the first American ecologist to think about fires in conifer forests. In the Heinselman Papers in the Minnesota Historical Society Library, there is a folder titled, in Bud's handwriting, "Four Historic Documents, 1910–1923."[13] One of these is a reprint of Frederick Clements's study of the fire history of forests composed largely of lodgepole pine in Estes Park in the Colorado Rocky Mountains.[14] As you'll recall from chapter 4, Clements was a pioneering American ecologist whose work on the succession of plant species after fires and other disturbances laid the groundwork for many ecologists after him, including Heinselman. Clements argued that disturbances that kill plants, such as fire, reset the plant community to an earlier stage of succession. After a large disturbance, Clements thought that plant communities will succeed one another in an orderly sequence culminating in a single, stable plant community characteristic of the regional climate.[15]

In order to more closely understand how fires affect forests, Clements reconstructed the fire history of the Estes Park lodgepole pine forests by taking cores or wedges from trees and counting the rings. He noted that some of these rings were charred, obviously the result of past fires that did not kill the trees. The live trees that had charred rings, therefore, preserved records of fires in the landscape. Inside the back cover of his copy of Clements's paper, Heinselman noted the years that Clements determined that fire swept through these forests: 1905, 1903, 1901, 1896, 1891, 1878, 1872, 1864, 1842, 1781, 1753, 1722, 1707. Every decade or two for two centuries, fires ignited the accumulated resinous needles on the forest floor and charred the trees at least once.

Clements's studies of lodgepole pine forests led him to modify his theory of climax plant communities. Through the studies, he realized that disturbances were not always random events caused by some process external to the plant community. Instead, traits of the pines determined whether the fires happened in the first place and how frequently, and allowed the lodgepole pines to regenerate themselves after the burn. The flammable and resinous lodgepole pine needles on the forest floor made fires more likely.

In addition, the fires enabled the next generation of pines to occupy the site because the pines themselves were adapted to the fires. In particular, their pinecones had thick resin coatings that kept them tightly shut after they dropped. Over the years, many cones accumulated on the forest floor. When the dry needles were ignited, the resins melted and the cones opened, releasing their seeds, which then germinated and carpeted the ground with the next generation of seedlings, a trait known as serotiny.[16] Clements realized that flammable needle litter fed the periodic fires in lodgepole pine forests, but the serotinous cones sustained the lodgepole pine populations after fire swept through. This was the first recognition that conifers not only might control the fire regime but were also adapted to it. Plant traits could both initiate fire and regenerate the same species after the fire had passed, thereby stopping succession at an earlier stage than the supposedly stable climax community.

Clements's paper inspired Heinselman to see if there was a similar fire cycle in the pine forests of the Boundary Waters. While using Clements's techniques of reconstructing the fire cycle from charred tree rings, Heinselman also added his own technique, mapping forests of different ages by the nature of their canopies. This combination allowed Heinselman to extend Clements's thinking about fires from just one species (lodgepole pine) to many other species, including white and red pine.

~

Although scientists try to describe their research methods in detail in their papers so that others can check their findings, it is always best to visit scientists in their laboratories or at their field sites to see firsthand how they do things. Such opportunities are rare, however. So, in late

October 1991, I and a number of Bud's colleagues were fortunate to be invited to a fire ecology symposium organized by the Friends of the Boundary Waters Canoe Area Wilderness. The highlight of this symposium was a canoe trip led by Bud to some of the oldest pine stands, where he obtained the longest record of fire history in the region.

From our canoes, Bud pointed out where the appearance of the canopy of the predominantly white and red pine forests changed abruptly at various points along the shore. This abrupt change suggested that the stands on either side of these boundaries began growing at different times. Younger forests have largely uniform canopies because they are composed of trees of approximately the same age but, as the forest ages, some trees die, others grow into the gaps they left, multiple episodes of ice, snow, and wind break branches, and the crowns and the entire canopy assume a rougher appearance.

With a little practice, we could easily identify forests of different ages while we canoed past them, just as Bud had done. Bud then beached his canoe on shore and sought out one or more of the largest pines, preferably one with a fire scar. Within the scar, the cambium (the living layers of bark and wood), which is only a few rings wide, has often been killed by fire but remains unharmed and alive around the rest of the circumference of the trunk. In subsequent years, this unharmed cambium begins to grow laterally into the scar. Eventually, the scar heals over completely from the two sides, leaving a telltale flat, triangular shape on the otherwise rounded trunk. Bud would take a core from the unscarred portion of the trunk to the center of the tree, sometimes cutting a small wedge from the healed fire scar with a handsaw. (The core and wedge do not harm the tree; many of the sampled trees remain standing and healthy.) He could use this wedge to date blackened rings from past fire(s), separated by normal ring growth. From these cores and wedges, Bud could obtain a fire history of the stand, often going back several centuries for very old sentinels of the forest.

Once he learned the age of a stand and its fire history, Bud then matched them to the appearance of the canopy and made field sketch maps of each stand from the canoe, noting its age and the years it burned. These field notes and maps would later be augmented with data from air photos and transferred to US Geological Survey topographic quadrangles. From these topographic maps, Bud could calculate the proportion of the total area of the quadrangle burned in each fire.[17] These annotated topographic maps helped him think through his many ideas about the role of fire in these forests.

When completed, these maps documented the scale and timing of major fires in this region since 1595, the age of the oldest pine trees that Heinselman cored.[18] The results were astounding, especially since they came at a time when Smokey Bear was telling us that forest fires destroyed forests forever. Approximately every twenty-six years in the Boundary Waters, enormous fires burned ten thousand to a hundred thousand acres. The cores and wedges Bud obtained from fire-scarred pines demonstrated that these fires happened in 1681, 1692, 1727, 1755–1759, 1801, 1824, 1863–1864, 1875, and 1894, similar to what Clements had seen in the lodgepole pine forests of the Rocky Mountains.[19] These fires were almost certainly caused by lightning, especially in drought years such as 1863–1864. After 1894, the fire suppression policies of the Forest Service effectively ended this natural fire cycle.

From these data, Heinselman defined a new term, the *natural fire rotation,* or the amount of time it takes to burn across an entire landscape. The natural fire rotation for most of the Boundary Waters forest is on average approximately one hundred years.[20] Although Heinselman found that every one of the forests in the million-acre Boundary Waters Canoe Area Wilderness originated after a fire, a natural fire rotation of a hundred years across the whole Wilderness doesn't mean that every forest never gets to be more than a century old. Some forests are burned more frequently, but some old forests survive periodic low-intensity ground fires. What happens in any one stand depends on the soil type and the tree species present.

At one extreme, for example, fires happen most frequently in jack pine or aspen–birch forests, especially on dry sandy soils formed in glacial outwash, where they recur every fifty to eighty years. Most jack pines are killed by fires, mainly because their lower branches, which are not self-pruned, carry fire from the ground up through the crown. But the cones of jack pine, as with cones of the lodgepole pine, are serotinous and remain tightly closed by resins until a fire opens them up and releases the seeds. The next year, the seeds germinate and the seedlings grow rapidly. Frequent fires sustain jack pine forests just as they sustain lodgepole pine forests. Aspen and birch are also easily killed by fire, but their roots remain viable because most fires do not heat the soil very deeply. The roots and stumps then sprout new stems that are the next generation.

The northern hardwood forests, which are dominated by sugar and red maple, yellow birch, and basswood, are at the opposite end of the range of fire regimes. These forests have a very long fire rotation, much longer than three hundred years. This is partly because the forests usually grow on the clay-rich soils of moraines, which can hold more moisture than sands on outwash plains occupied by the drought-tolerant jack pines. Unlike conifer forest floors, the flat leaves of maple, yellow birch, basswood, and beech decompose quickly, so fuel does not accumulate as it does with the more slowly decomposing jack pine needles. Their flat leaves also overlap one another on the forest floor like shingles on a roof, allowing films of water to become trapped between layers. Fires cannot get a purchase in these thinner and moister forest floors except during prolonged drought and abnormally warm weather. Because northern hardwood forests rarely burn, they are often called asbestos forests.

The fire regimes of white and red pine stands are more complex than either of these two extremes. The fire regimes of white and red pine–dominated forests consist of severe crown fires every 120 to 180 years, punctuated by more frequent but lighter ground fires every 30 to 50 years or so.[21] As in jack pine forests, white and red pine forest floors burn with ease because their long and resinous needles decompose slowly and many years of needle litter fuel accumulate.

As opposed to jack pine or aspen–birch, most large, old white and red pines survive ground fires quite well because trees older than forty years have very thick and corky bark at their bases that can protect the larger trees from even quite severe ground fires. In addition, white and red pines shed their lower branches as they age, so ground fires do not usually reach into their crowns. These pines that survive ground fires are the ones with scars at their bases but continue to produce good crops of cones every three to five years. The ground fires kill most thin-barked species of the understory such as maples, juneberries, viburnums, and other hardwoods. Consequently, white and red pine seedlings, germinating from the seeds produced by the old monarchs that survived the fire, are less shaded and become the next generation of trees in the now-cleared understory.

After long periods without ground fires, however, understory spruce and fir can grow up to the lower branches of the taller pines. These spruce and fir trees can carry a ground fire up through their ladder of branches into the overstory canopy of pines and ignite a crown fire. Unlike the sizzle of ground fires, crown fires roar like freight trains as they suck air upward into the canopy. Severe crown fires create their own winds that toss burning embers and branches ahead of the fire perimeter, creating spot fires that may also become crown fires. However, in even the largest and hottest crown fires, a few white or red pines usually survive in isolated pockets. These pockets of unburned trees reseed the surrounding burned area.

The periodic clearing of the understory by ground fires and the survival of at least a few old seed-producing trees after crown fires are probably the major ways red and white pine populations sustained themselves in the virgin forests of the Boundary Waters.[22] Were it not for this complex fire regime, both white and red pine would eventually succeed to the more shade-tolerant northern hardwoods.

In his classic book *Arctic Dreams*, Barry Lopez says that the complexities of northern ecosystems lie not with, say, "100 species of ground beetle in the same acre as in tropical forests, but with an intricacy of rhythmic response to extreme seasonal ranges of light and temperature."[23] Heinselman's work demonstrated that the northern pine-dominated forests have an intricate rhythmic complexity that is shaped by fires over the course of centuries. This complexity is shaped partly by the traits of some species that promote or suppress fires and other traits that enable the populations to recover from fires. This interplay between fires and species traits creates a landscape with fires of various return intervals, severities, and sizes and a diversity of species adapted to them. In 1970, ecologist Orie Loucks at the University of Wisconsin suggested that periodic fires may be necessary to maintain the diversity of temperate and northern forest ecosystems, most notably the pine-dominated North Woods.[24] This interplay turns Smokey Bear's message on its head, at least for natural wildfires set by lightning. Smokey Bear is correct that we should be careful with fire in the forest—but not all fires should be suppressed. An individual white pine tree or even a stand might be killed by fire, but the presence of fire sustains white pine forests across the landscape.

~

Heinselman's research led him to advocate for letting natural wildfires continue to maintain these pine forests, eventually resulting in the Superior National Forest's Fire Management Plan in 1991, one of the first plans to take natural wildfires into account in forest management.[25] The plan states that lightning-caused fires in the Boundary Waters will be allowed to play, as much as possible, their natural roles in the Wilderness Area *unless* there are significant threats to human life or private property; there is unacceptable drought; if fire behavior or weather will likely cause the fire to exceed managers' prescriptions for the fire (as if managers could prescribe what a fire should do); the impacts of smoke on nearby resorts and towns are unacceptable; if the fires pose unacceptable inconvenience, annoyance, or economic hardships to nearby communities; if the fire threatens to cross the wilderness boundary (as if the fire knows where the boundary is); and if there are not sufficient personnel to fight a larger fire. If any of these caveats hold, then the fire is suppressed.[26]

This rather circumscribed fire policy has been caricatured among some Forest Service personnel as, "The Forest Service has a let-it-burn policy until the fire starts, then we put it out." These policies remain in effect even today. Nonetheless, these caveats are not unreasonable, which is why fire management requires extreme courage and nerves of steel. Even as I write this in a hot and dry summer in northern Minnesota more than thirty years later, fires are burning throughout the Quetico–Superior Wilderness and just outside it in the Superior National Forest. Each fire is assessed and managed individually according to whether it was started by lightning and whether it poses threats outlined in the above list of caveats. Some are allowed to burn, some are contained by firebreaks, and some are actively suppressed.[27] This is a lesson to ecologists: just because we understand the ecology of a species or an ecosystem doesn't mean that making policy for it is simple.

~

Aldo Leopold wrote that the highest value of wilderness areas was not for recreation but instead as natural land laboratories in which we could discover how natural ecosystems worked.[28] Heinselman's studies of fire in the northern pine wilderness of the Quetico–Superior are superb examples of the importance of such natural land laboratories for learning how landscapes work. Only a large, contiguous wilderness area "untrammeled by man" such as the Boundary Waters could offer Heinselman a natural laboratory for uncovering how the complexity of fire regimes in northern conifer forests maintains a diversity of plant communities. A smaller area might burn entirely in one fire, without giving ecologists the chance to understand the relationship between fires, the diversity of species, and the stability of an ecosystem. In his research on the fire-and-conifer forests of northern Minnesota, Heinselman laid the groundwork for thinking about how large-scale processes like fire shape landscapes. He also showed how to put that thinking to the test, combining field data and maps to discover how today's forests emerged from the fires of the past. The quantitative analysis of maps pioneered by Heinselman is now a core procedure in the new field of landscape ecology.

Heinselman's research program began as a plan to learn about fires to "obtain a forest of the long-lived species, such as the red pine, the white pine, and the white spruce," as the Selke Committee recommended. What he learned went far beyond the recommendation to "obtain" such a forest. Heinselman showed that fires are necessary to sustain the forest, not simply obtain it.

In the evening after our canoe trip with Bud, we discussed some of the questions his research prompted that remained unanswered (and remain so to this day). The most important question, which Bud said he could not answer and in some ways we still do not understand, was "How does a fire spread through a landscape that previous fires have created?" As the climate warms and fires become widespread in the pine forests of the Boundary Waters and throughout much of the American West, answering this question becomes more urgent for not only managing wilderness areas but also managing and sustaining forests outside the wilderness. Once we better understand the processes that sustain a natural ecosystem, we can apply that knowledge to manage areas outside the wilderness. However, the large sizes (100,000 acres or more) and long recurrence intervals (within one or two human lifetimes) of these natural fires make it impossible to maintain such complex natural fire regimes outside the wilderness, even in such sparsely populated areas as northern Minnesota. To sustain white pine as fires have done in the past, we will have to apply what we have learned using other forest management techniques that mimic the effects of fire. But doing that often requires several generations of both people and trees.

CHAPTER 10

Restoring the White Pine

In 1902, Ivar Rajala (pronounced *Rye-la*) and his wife, Anna, left their home in Kankaanpää, Finland, and joined thousands of other Finns headed for the Lake Superior region, where they settled on a homestead near Effie in northern Minnesota. There, Anna gave birth to seven sons and one daughter, and she and Ivar cleared the land and built a farm.[1] The heyday of white pine logging in northern Minnesota was just passing, and Weyerhauser and other companies were moving on to the Pacific Northwest. But isolated white and red pine stands and individual large trees remained north of the divide between the Mississippi River Watershed and the Hudson Bay Watershed. Ivar and Anna realized that, while the homestead would supply them with food, the pines would supply them with income. So Ivar and his sons began contract logging for the Minnesota and Ontario Paper Company, sending pulpwood to the paper mill in International Falls and larger white pine sawlogs to the few remaining sawmills in the area. Two-man saws, heavy axes, oxen, and horses were still used to fell the trees and move the logs out of the woods. Loggers lived in camps in the woods in winter and ran the logs down the rivers in spring. The Rajala Camp was one of the last logging camps in the North Woods and marked the end of the Paul Bunyan era.

As fate would have it, John Larson, who had immigrated from Sweden, built a sawmill on the banks of the Big Fork River about six miles south of Effie in 1902, the same year Ivar and Anna left Finland. Larson bought his main band saw and the carriage that shuttled the logs through it from Henry Ford, who had used them to saw white pine in Michigan at the same time he was making the Model T. Logs were driven down the Big Fork River to the mill pond, where they were stored until they could be hauled out of the water and into the mill. Ivar eventually bought the mill from Larson and thereby started the Rajala legacy of foresters, loggers, and sawyers—now in its fifth generation. The Rajala mill in the town of Bigfork, Minnesota, is now the oldest continuously operating sawmill in the Lake States.

Ivar's sons later purchased a twenty-three-thousand-acre expanse of prime timberland, known as the Wolf Lake Tract. In previous decades, some pines in the Wolf Lake Tract had been harvested by the Joyce, Hartley, and Pillsbury families (yes, that Pillsbury). But for the most part, these families and their friends used it as their private hunting and fishing camp. When the Rajalas bought the Wolf Lake Tract, much of its forests were near-pristine North Woods. The Rajalas knew they had a jewel of a forest that, with some restoration of white pine, could supply large logs for their sawmill for many years. The journey of the Rajala family toward that goal embodies the changes in silviculture since the plantation forestry of the Civilian Conservation Corps.

Beginning in the 1960s, the surrounding Chippewa National Forest and various state and industrial forests were clear-cut every forty to fifty years in order to supply smaller aspen pulpwood for the region's papermills. Isolated large white and red pines from these surrounding forests made their way to the Rajala mill, but the supply was dwindling. In short, the remaining white and red pines in the surrounding forests were being liquidated while the more rapidly growing aspen was being

managed to supply regional papermills. Ivar's grandson, Jack Rajala, and his son, John, realized that their management of the Wolf Lake Tract for large trees and beautiful lumber, for which their family was known, was being subsidized by the unsustainable harvest of large pines in the surrounding public forests. Something had to change.

Jack became a champion for restoring white pine to its former dominance throughout northern Minnesota. He supported both preserving wilderness areas and restoring a white pine sawtimber industry along with the pulpwood industry that had come to dominate northern Minnesota. At this time, decades of research and timber harvest had formed the tenets of classical silviculture, through which foresters managed a single species based on its biology. This is what Gifford Pinchot, Volney Spalding, and Bernard Fernow wrote about at the beginning of the twentieth century and what had been taught in forestry schools. Jack combined this classical approach to silviculture with his own lifetime of experience in a handbook called *Bringing Back the White Pine*.[2] As much as possible, Jack promoted natural regeneration of white pine from seeds produced by older trees temporarily protected from harvesting.

When John took over the business after Jack's untimely death, he realized that it would take longer for these seedlings to grow into the large sizes needed for quality sawtimber than the remaining supply of their large white pines would last. Bringing the white pine population into a state where the growth of younger trees could keep pace with the harvest of older trees would take some decades to achieve. In the meantime, to supply customers with a variety of quality lumber, the company also began active management of hardwoods such as maple, birch, and oak to supplement white pine. The sale of high-quality products would pay for the land management required to bring the forests back to their former glory and sustain the harvest of timber from them. The management of the full complement of species in the Wolf Lake Tract, indeed the entire ecosystem, demanded sustainable management of almost all its tree species, especially the long-lived maples, yellow birches, and red oaks in addition to white and red pines. If it could be done, both the company and the forest would benefit.

Fortunately, silviculture itself was moving in a direction that would provide John with the tools to do just that. While Jack practiced classical silviculture, John, his son, has become one of the earliest private practitioners of what is now known as ecological silviculture. Ecological silviculture had its origins in the many conservation battles of the late twentieth century, when foresters and forest ecologists realized that focusing on a single timber species was leading to a loss of habitat and species diversity required to sustain productivity.[3] Foresters became increasingly aware that a timber species depended on the rest of the ecosystem, including processes like fires, to obtain nutrients, pollinate cones, disperse seeds, provide openings for seedlings, and for other aspects of its life cycle. Forestry had to move away from single-species management and instead manage for more complex ecosystems that would, in the words of Brian Palik's and colleagues' current bible of ecological silviculture: "Emulate natural disturbances, sustain biological legacies, emulate developmental processes, and allow time to take its course in shaping stands, all to the benefit of ecosystems and the people who use them."[4]

That, at least, is the goal John Rajala set for his company.

~

If you walk through the Wolf Lake Tract on a beautiful autumn day and stand on the shore of one of its several small lakes, you will see a classic North Woods scene. Large pines, maples, and red oaks surround you, along with saplings and smaller trees in the understory. Along a moraine on the opposite shore of the lake, even more large pines thrust their crowns through a blazing canopy of sugar and red maples, yellow birches, red oak, and possibly a basswood or two, punctuated by a few white spruce. Two or three of the fifty or so bald eagles that nest in the crowns of white pines on the tract might circle overhead, getting ready to migrate south as the lake starts to freeze. You might think you are on a late-season canoe trip through the Boundary Waters Canoe Area Wilderness to the north. But a closer look reveals a stump here and a stump there, often from a recently cut tree. In places, the understory and some of the overstory are being cleared, leaving two or three of the larger trees on each acre, much as if a fire had recently burned through a patch of the forest. But it wasn't a fire that thinned the forest. Rather, it was a logger who felled the trees with chain saw or feller buncher, then skidded the logs to a landing, where they were cut into sixteen-foot lengths (or sometimes longer for special orders) and transported by truck to the Bigfork sawmill.

The objective of Rajala's ecological silviculture is to use logging to mimic fire, maintaining and restoring the white pine along with a rich diversity of other North Woods tree species, every one of which can be turned into valuable lumber if allowed to grow to large sizes. This is not so easy. Just as white pine in the wilderness requires a complex fire regime to sustain itself, maintaining white pine in a managed forest requires what forest ecologists call "a finely tuned balance between natural disturbance and management action."[5]

Mid-successional forests, where pines tower over northern hardwoods, will not stay in this state indefinitely, because the seedlings of the pines cannot grow as abundantly in the shade of the understory as the seedlings of the more shade-tolerant northern hardwoods. In natural forests, white pine seedlings and saplings do best where canopy openings allow more than 35 percent of full sunlight to reach the forest floor and where there is little competition from deciduous hardwood species.[6] As we saw in chapter 9, these conditions were frequently created by ground fires that also killed the occasional large tree.

Can logging mimic these conditions? Both logging and fire open up the canopy to various degrees, but they affect the ecosystem and seedling and sapling survival differently. Fires, especially ground fires, remove shrubs while leaving the big mature seed trees behind, although some are killed. In order to make money, loggers have to take the big trees, often leaving the shrubs and commercially less valuable small trees behind. There is often less shrub cover on burned sites than on logged sites, and so pine seedlings are not so heavily shaded; consequently, more white pine seedlings reach sapling and larger sizes on burned sites than on logged sites.[7] To mimic the better conditions for seedlings after a fire, John Rajala therefore instructs his loggers to "brush out," or remove, many shrubs and smaller saplings except for the ones that are straight and healthy.

Much of John's thinking, however, concerns what to do with the big trees, not only white pines but all other species. Whether or not he decides to cut a particular large and valuable tree for the mill depends partly on the quality of the logs that this tree would yield. This consideration has to be balanced against the ecological functions the tree provides. Walking through the forest with John, I learned that he asks himself five questions when deciding what to cut and what to leave:

1. Is the tree of the size and quality needed to yield enough lumber to pay for cutting and milling it?
2. Are there enough seedlings and saplings to ensure that at least one will replace the tree to be cut?
3. If not, can other neighboring trees provide good seed crops?
4. Is the tree interfering with the growth of a neighboring, more desirable tree, or with seedlings and saplings that will form the next generation?
5. Could other nearby trees also provide the habitat this tree provides for other organisms essential to the ecosystem (e.g., mycorrhizal

symbioses; voles to disperse mycorrhizal spores; red squirrels to disperse seeds; nest sites for predators that keep voles, squirrels, and insects in check)?

If the answer to all these questions is yes, then the tree can be cut without reducing the sustainability of the forest. If the answer to at least one question is no, then the ecological and economic value the tree could provide has to be weighed against the answers to the other questions.

One of the most important steps an ecological forester can take today is protecting white pine seedlings and saplings from their greatest threat: white-tailed deer. White-tailed deer are far more abundant throughout eastern North America today than before European settlement. Because of that, deer are a serious issue in forest management throughout eastern North America, not just in white pine forests.[8] Deer browse the apical (terminal) shoots of white pines and other species, which inhibits the growth of these shoots and allows the seedlings to be overtopped by their neighbors, especially ironwood (*Ostrya virginiana*), which deer do not browse and which is of no commercial value. The unbrowsed ironwood can shade shorter white pine seedlings and suppress their growth. As John says, the pines are in a race with their competitors and with the deer. A high deer population and healthy white pine forests are simply not mutually sustainable—a point that is lost on the general public and often on wildlife managers. Because it takes many years for seedlings to become established and grow to the height where deer can no longer browse them, the effect of deer on the composition of the forest can persist for several decades. Merely reducing the deer population by hunting, therefore, will not immediately solve the problem.[9]

To protect the growing seedlings and saplings, John and many other landowners cap the trees' terminal buds in early fall when deer start browsing. Caps are made by folding a piece of recycled paper around the bud and stapling the ends together.[10] Not every seedling or sapling has to be bud-capped, just the ones of superior growth and form. Bud capping requires great effort to preserve promising young trees for which the financial returns are deferred well into the future. Once a white pine is over six feet tall, its apical shoot is beyond the reach of deer, and the sapling has a good chance of surviving competition from neighbors, so bud capping is no longer necessary.

These complex decisions—which trees to cut and which to leave, how to open the canopy, and which seedlings or saplings to protect from deer—have to be made for each of the commercially valuable species in the Wolf Lake Tract. Sustaining a valuable mixed-species forest like this means that John Rajala must make these decisions on almost a tree-by-tree basis.[11] The complexity of the forest makes it harder to manage, but it also provides a species-diverse and beautiful landscape that is a mosaic of stands and trees of different ages and species composition, including very old trees, from which a great variety of valuable timbers are harvested and trucked to the mills.

Henry Ford's band saw and carriage still mill logs of white pine, maple, red oak, birch, and other species from Rajala timberlands. The steel that was used to make these saw blades is of very high quality, and it is difficult to find modern blades that could keep the sharp edge of the old saws. Just as in the old logging camps, the mill has a full-time filer who sharpens the teeth of the saws. Today, though, the old band saw and carriage are aided by computer controls and lasers that guide the logs to the blades. Slabs and sawdust are burned in a boiler to provide heat for the buildings and a kiln drier while excess sawdust and wood chips are sold for poultry bedding. The entire sawmill operation is a combination of the best machinery from Henry Ford's day with new technology that increases precision sawing and reduces energy demands and waste.

At one time, the Rajala's mills produced lumber of all sizes and grades, from construction-grade two-by-fours and two-by-sixes for framing houses to fine-paneling and trim for interiors. Today, it is difficult for a small sawmill to compete with big-box lumberyards that import pine lumber from overseas, so the Rajala family no longer mills construction lumber.[12] Instead, the Rajala companies concentrate on producing high-quality lumber from maple, birch, oak, basswood, and other species as well as the companies' signature white pines. Its storehouse contains one of the greatest varieties of species and grades of lumber anywhere in the Lake States. Some stands in the Wolf Lake Tract yield exceptionally distinct logs. The saw sometimes uncovers a birch that has a slightly reddish sheen or a maple that has a rich brown heartwood, which John markets as "red birch" and "brown maple," respectively. An advantage of vertically integrated companies like the Rajalas' that harvest and process wood from the forest through the mill and deliver it directly to the customer is that they know where in the woods the lumber came from. Fine timber, like fine wines, has a terroir. With attention, the Rajala companies might be able to manage their forests for a sustainable supply of unique specimens of lumber. In contrast to this vertical integration in the Rajala companies, logs harvested from lands in many ownerships bought by a large mill are mixed together, and the provenance, or origin, of these logs is lost.

Unfortunately, few timber companies have a forest in as good a condition as the Wolf Lake Tract and the sawmills to turn it into high-quality lumber. The Menominee Reservation in northeastern Wisconsin is an interesting exception, with an approach that is similar to that of the Rajala company in many ways. The Menominee tribe is a member of the Algonquian nation within the Anishinaabeg peoples. At over 350,000 square miles, the Menominee Reservation is the largest reservation east of the Mississippi River. The forests of the Menominee Reservation have never been clear-cut and have a diversity of species and older-aged trees that is broadly similar to the Wolf Lake Tract. Like the Rajala companies, the reservation both manages the forest and operates its own sawmill. Management is guided by the Menominee belief that humans and nature are not separate, that it is proper to harvest or hunt when the dignity of other organisms is respected, and that the entire forest has spiritual power.

From these spiritual beliefs, the Menominee have arrived at a set of management practices that mirror and long predate the principles of ecological silviculture. They include making the long-term benefit of the forest a higher priority than supplying timber to the sawmill, selectively harvesting individual trees, maintaining a high-diversity forest containing trees of all ages, and waiting until trees are at least two hundred years old before harvesting. It may seem that all these principles would decrease the amount of timber harvested from the forest, but instead the annual cut has increased from twenty million board feet at the start of the twentieth century to thirty million board feet today. This increased harvest is because fewer but very old and large trees are cut instead of many smaller trees harvested on shorter rotations, as in many industrial forests.[13] Waiting until trees are large, old, and valuable before harvesting is what both ecological silviculture and tribal spiritual beliefs both recommend. The irony is that these practices are not only ecologically and spiritually sound, they also improve the forest and make money at the same time.

Ecological silviculture is most easily implemented when the forest has not already been heavily degraded, such as in the Rajala family's Wolf Lake Tract and on the Menominee Reservation. Ecological silviculture is being practiced in various national forests, some of which include white and red pine forests.[14] But unless a company already has access to a forest in a reasonably good condition, ecological silviculture may be easiest to practice in public forests that are subsidized by public monies.

In the past, the Rajala companies' lumber was loaded on railcars in nearby Grand Rapids, hauled to the Twin Ports of Duluth, Minnesota, and Superior, Wisconsin, and shipped worldwide. However, John began to wonder whether the carbon released from burning fossil fuel to transport lumber far outweighed the carbon stored in the wood. He has ended worldwide shipping for now, instead concentrating on marketing quality lumber to regional carpenters, furniture makers, and homeowners so as not to incur the carbon cost of distant shipping. On a day I spent with him, the millworkers were preparing a shipment to southern Minnesota, and John was wondering what the carbon cost of even this relatively short transport would be.

If handled properly, a managed forest can remove and store large amounts of carbon dioxide from the atmosphere. Half the dry weight of a tree, indeed of most organisms, is carbon. In a tree, most of that carbon is stored in the wood. The longer the tree remains alive and standing in a forest, the more carbon it stores.

But that is only half the carbon balance. Carbon is stored not only in the growing stock of trees and the soil in the forest but also in harvested products as they move through the economy and generate revenue. Half the weight of any piece of wood in your home's two-by-four studs, siding, paneling, trim, flooring, and furniture is carbon. Your house and furniture are carbon storage devices that will last for many decades. In contrast, the lifetime of a piece of paper is only a few years. Once a piece of paper decomposes in a landfill or is burned in an incinerator, the carbon it contains is released back into the atmosphere. The longer the life of the tree before it is harvested and the longer the life of the harvested product, the more carbon is stored.[15]

A timber industry that manages for large trees on long rotations and then stores the carbon in harvested products with long lifetimes maximizes carbon storage even if the trees are so old that their growth is slowing. (See the discussion in chapter 6 about how the growth of a tree changes as it ages.) This is exactly what the Rajala companies are doing. A timber industry that manages forests on short rotations for short-lived paper products stores less carbon even though the forest is rapidly growing compared with a forest managed on long rotations for large but slowly growing trees turned into long-lived lumber products.

But there is an interesting twist in John Rajala's Wolf Lake Tract, where there are a few rapidly growing aspen and birch stands from a few small clear-cuts done many years ago. Rather than harvest these stands on short rotations for pulpwood that will be made into paper, John is still managing them on longer rotations, albeit not nearly so long as white pine, maple, and oak. When the aspen and birch are harvested, they too are sawn and milled into lumber that is sold for paneling, doors, and occasionally furniture, so the carbon in their lumber is still being stored while it makes its way through the economy. In fact, these species are performing an important role of rapid carbon removal from the atmosphere and storage in wood while the white pines and other long-lived trees await the time of their harvest.

One of John Rajala's goals is to manage his forest and produce long-lived products that will make his company a net carbon sink. The forest management and milling practices that John Rajala implements offer a promising pathway for storing carbon and keeping it from the atmosphere for many decades, even centuries. This is an example of what the Nature Conservancy has termed natural climate solutions. Of course, this type of timber industry cannot solve the global climate issue on its own. In the United States, such forest management practices can store only approximately 20 percent of what is needed to keep warming to less than 1.5 degrees Celsius, the goal of the Paris Agreement and other climate accords.[16] Nonetheless, it is an important 20 percent, which can augment other needed solutions, such as decreasing dependence on fossil fuels. Approaches like the Rajala family's, which care for forest health and carbon capture as well as economics, will be ever more necessary as climate change accelerates threats to white pine.

CHAPTER 11

Climate Change and the Future of White Pine

At the beginning of this book, we followed the genus *Pinus* as it diversified, expanded, contracted, grew, and shifted in response to many millennia of climate change. Eighteen thousand years ago, during the Last Glacial Maximum, land temperatures in mid-latitudes, where eastern white pine was then located, were about ten to twelve Fahrenheit degrees cooler than today.[1] As we saw in chapter 1, the retreat of the ice sheet and the rise in temperature slowly drove white pine northward and westward until, about two thousand years ago, it reached its current distribution. The mean annual temperature in the middle of the range of white pine today is approximately 40°F. If the concentrations of carbon dioxide and other greenhouse gases continue to rise at their current rates over the next century, mean annual temperatures in the same region could reach 50–60°F, meeting or exceeding temperatures now experienced in many places along white pine's southern border. Clearly, the future distribution of white pine will not be the same as it is now.[2] The changes in temperature are too large for white pine to remain in many places it is found today.

As climate change progresses as anticipated, where will white pine be found in the future, and how will it evolve? We don't precisely know. Various models use the responses of the North Woods and white pine to current and past climates to predict what will happen in future climates.[3] The climatic factors that correspond with a species' range today are collectivity known as the species' climate envelope. The models predict where the species' climate envelope will be under different climate change scenarios and assume that the species' population will migrate in pace with its envelope. Most of these models predict a general northward and perhaps slightly eastward shift of white pine throughout most of its current range in the United States. Under the largest predicted warming, white pine in the United States may remain common only in northeastern Minnesota and northern Maine, although less abundant than today. The potential center of white pine's range will likely extend well into what is now occupied by boreal spruce–fir forest in Ontario and Quebec north of Lakes Superior and Huron.[4] Today, white pine is less common or even absent throughout much of this area.

Plants, of course, do not "migrate" like birds, caribou, and other animals. Instead, ranges shift as seeds are dispersed to areas outside the current borders; only those that land in favorable sites can germinate and grow (assuming they are not browsed by deer or killed by insects, disease, or pests) until they are old enough to make new seeds, which are then dispersed further. White pine generally does not produce abundant cones until it is about twenty to thirty years old, and even then, abundant cone crops are produced only every three to five years, so this "hopscotch" shift in its range will be patchy, episodic, and slow.

During the eighteen thousand years since the ice sheet retreated, most tree species of eastern North America, including white pine, migrated at an average rate of twenty to forty miles per century as the climate warmed. The nearly ten-degree rise in temperature that some models predict white pine could experience over the next hundred years would be nearly the same amount that the temperature rose in the eighteen thousand years since deglaciation. This rapid projected rate of climate range will require that species migrate at least ten times faster than in the past to stay abreast of its climate envelope.[5]

In addition, natural barriers such as Lakes Superior and Huron would almost certainly block white pine seeds from populations in Wisconsin, Michigan, and Western New York from dispersing to Canada. (Birds such as crossbills and redpolls, which eat pine seeds, generally do not fly across these Great Lakes, and squirrels certainly cannot swim across them.) In additional to these natural barriers in the center of white pine's range, vast swaths of cities and suburban areas, highways, and other structures of what is now called "the built environment" could also impede migration. The populations of white pine in the southern half of its range may be "preadapted" to the warmer climate anticipated to develop north of its current range, but the warmer climate might move out from under them faster than they can disperse northward around these barriers.

If white pine cannot disperse fast enough to keep pace with its climate envelope, can it evolve fast enough to adapt to the warmer climates expected within its current range? During the long, slow migration north and west out of its Ice Age refuge in North Carolina, white pine evolved to help the trees survive the shorter northern summers and long and snowier northern winters, as we saw in chapter 1. So, biologically significant evolutionary changes did sometimes keep pace with the slower rates of past climate changes.[6] It is unlikely, however, that traits will evolve fast enough to keep pace with more rapid future climate change.[7]

Other factors would also impact white pine survival. As the scope of its natural history and evolution shows us, white pine is intricately bound into a web of communities, which include other plants, animals, and fungi. Whether the mycorrhizae, squirrels, and other species upon which white pine's life cycle depends will migrate along with white pine remains unknown. Species do not exist in a vacuum but are knitted together by the flow of energy and nutrients in foods webs and ecosystems. But food webs and ecosystems do not migrate whole. Individual species do, and not always at the same time and in the same directions. The ecosystem that white pine both dominates and depends upon may not remain intact during the warmer climate of the next century. New connections with other species may have to evolve in order for white pine populations to be sustainable, but this takes time.

So is white pine doomed to extinction? Probably not, but where it is found and how abundant it is will certainly change during the next century, although we cannot accurately predict what will happen yet.

Perhaps there are things that we can do to help white pine as well as other species cope with the climate warming that has begun. Certainly, anything that reduces the carbon dioxide burden of the atmosphere—reductions in fossil fuel use and deforestation and, as we saw in chapter 10, expanding long-rotation forestry and milling durable timber products—will slow climate change, perhaps even to the point where species migrations and adaptations may be able to keep pace.

But if southern populations cannot migrate north fast enough, we may be able to help them along and assist their migration or colonization. We can collect seeds from southern "warm-adapted" populations and then plant them or their seedlings where the southern climate is expected to be in the coming decades.[8] For example, white pine seeds or seedlings collected from Southern Ontario might be suitable for planting along the current northern third of white pine's range as the climate warms.[9] Models can be used to help guide such decisions, but large-scale field trials using seedlings collected from southern and northern sources would be a better, albeit an expensive, way to identify the best places to obtain seeds and seedlings. It may take decades, however, to collect enough data from a field trial to pinpoint the correct seed sources: the southern populations that will grow best in a northern location when they are mature decades from now may not grow the best there today.

Although the work will be challenging, the Nature Conservancy has already begun such long-term field trials, including with white pine, in northern Minnesota.[10]

Assisted migration or colonization has critics who point out its risks, such as unintended spread of disease or pests from the source site to the colonized sites and the chance that the transplanted population could become invasive in its new habitat; there are also bioethical issues, such as how to balance the preservation of a commercially valuable species such as white pine against the benefits and risks to the new ecosystem to which it will be transplanted.[11] The history of white pine is certainly full of cautionary tales when it comes to humans' attempts to repair forests, such as the spread of blister rust to America from infected European seedlings. Others suggest that these risks may be low if the managed shifts in populations are gradual.[12]

Northern conifers, including white pine, may also survive in cooler refuges in the landscape. In chapter 1, we learned about Constance Millar, a conifer biologist with the Forest Service who proposed that ancestors of today's North American pines survived warmer periods during the Eocene by retreating into high-elevation refuges in the Rocky Mountains. Building on this hypothesis, Millar and colleagues have proposed that conservation biologists identify and experiment with environmental refugia for species that otherwise may not survive warmer regional temperatures and perhaps drier conditions.[13] Scientists with the Nature Conservancy found that easily mapped features of the landscape could be used to locate places where various species could survive in a changing climate.[14]

For example, low spots or north-facing slopes that are cool and moist might serve as refuges for white pine and other northern conifers.[15] Termed conifer strongholds by the Minnesota–South Dakota–North Dakota chapter of the Nature Conservancy, these sites may become foci for reseeding and recolonization of the surrounding areas when the climate begins to cool.[16] Protecting or planting conifers on these sites would increase the resiliency of the forest by enabling it to recover from climate change.

To implement a trial of conifer strongholds, the Nature Conservancy purchased 158,000 seedlings, consisting of 28,000 white pine and white spruce from sources in southern Minnesota and 130,000 seeds of white and red pine, white and black spruce, jack pine, tamarack, and northern white cedar from local sources in northern Minnesota. These seeds were sown in a nursery, and in 2017 and 2018, the two-year-old seedlings were planted across 1,169 acres of federal, state, and county forestland in northeastern Minnesota. Scientists chose sites representing all combinations of warm and cool temperatures and high and low topographic diversity for planting. They also chose an additional set of unplanted sites to determine the abundance and success of natural seed production and seedling establishment. The Nature Conservancy will monitor these sites every two years to assess the feasibility of conifer strongholds for preserving white pine and northern conifers in a warming world.

These vast numbers emphasize how large even experimental tests of these ideas need to be in order to tease apart the various environmental factors at work. It is imperative that similar trials be established elsewhere for these and other species. The knowledge we gain from such experiments will be invaluable for the conservation of these species and the food webs they support.

~

As we look forward to the future of white pine, let us return to the old-growth white pine forest we first visited in chapter 1, which lies within Superior National Forest near Duluth, Minnesota. As we walk under the pines, each of us is in the same forest; we each hear the hissing of the wind through the needles high above, the scolding of the red squirrel, and the whacking of the pileated woodpecker against the trunk; we each drink in the cool air perfumed with the tang of pine resins. It is the same for all of us, but not the same. The meanings we give to these impressions depend on which stories each of us values about white pine: its spiritual nature; our ability to safely fell the trees and turn them into beautiful lumber; the feeling of awe that arises in a wilderness setting, as if we are present at creation; our understanding of the natural history and ecology of white pine and the other species it lives with; and whether and how white pine is part of our family histories. Our stories make this forest special—just as they do other wild places. These sometimes-conflicting stories did not replace one another during the history of humans' association with white pine. Instead, they were layered upon one another, and each of us probably accepts several of them.

And therein lie the conflicts between us about what we should do with this forest. We prefer our own personal story, but none are intrinsically superior. The richness of the cultures of a place, any place, derive from the variety and complexity of stories woven around and through it. In turn, each one of us is defined by those places and through the stories we tell about them. These stories often center on a characteristic species: Arizonans are the people of the saguaro cactus; Pacific Northwesters are the people of the Douglas fir; Vermonters are the people of the sugar maple; and Minnesotans, Iroquois, and Algonquins are the peoples of the white pine. Such species become a foundation for how resident populations view the land and their obligations, if any, to it. Often, these stories can coexist even if they are in conflict. But when one narrative begins to dominate all others or when the foundation species declines, then the conflicts become real and seem difficult, perhaps impossible, to resolve.

If, however, we can merge the different views into a coherent new and richer story, then we may have the tools to resolve our conflicts. Something of the sort may be happening today for white pine as the practice of scientific ecological silviculture converges with similar approaches based on millennia-old spiritual understandings of our obligations to a forest, as practiced by the Menominee in Wisconsin, for example. The accelerating changes in the climate that all ecosystems will experience in the coming decades give an urgency to reconciling these conflicts.

What should be done with this particular white pine forest? Should we just let nature take its course and leave it alone? This forest, which is in a rather cold place even for northern Minnesota, may survive some amount of warming and may even thrive as warmer temperatures increase the growth of monarch pines and the survival of their seedlings. If white pine seeds disperse into surrounding areas while spruce and fir decline, this forest may even become a conifer stronghold, much as white pine temporarily replaced spruce and fir during a warmer climate approximately ten thousand years ago. The current generation of white pines might be integral to maintaining the larger forest's resilience to climate warming.

A warmer and potentially drier climate in this region, however, brings with it a higher probability of fire. In the summer of 2021, a particularly hot and dry season, a wildfire burned nearly twenty-seven thousand acres of similar forest a few miles north of this one. If a fire starts here in the future, should we let it burn and allow the forest to regenerate itself naturally, as similar forests did for centuries in the Boundary Waters Wilderness Canoe Area not far away? Should we put it out and "save" this forest? Or should we start a prescribed ground fire and mimic nature in a more controlled way? In all these options, the fire will put carbon dioxide stored in the wood back into the atmosphere. While this forest will generate only a tiny amount of carbon dioxide, the cumulative impact of many other similar forests burning is much greater. Many of the trees in this forest are large and have clear boles that could provide beautiful lumber. If the forest is going to burn anyhow, as it likely will at some point, should we harvest at least some of the trees, preserving the beauty of their wood in houses and furniture that will continue to store carbon?

If, however, these particular white pines will die in a warmer climate, should we plant seedlings from a more southerly location that may thrive here? That might preserve a pine forest here, although not of the same genetic stock. Should we collect the seeds from these trees and transplant them northward to where today's climate might be in the future? That might preserve this genetic stock, but it will be living elsewhere, perhaps in combination with a new set of species in a very different future forest. Could this genetic stock be successful in a different kind of forest? The success of these white pine transplants and whether they form the foundations of new ecosystems will depend in part on whether they can make connections with all the other species they depend on and that depend on them. These include the mycorrhizae in the soil; the squirrels that disperse the white pine seeds; the woodpeckers that eat the carpenter ants, protecting the trees while receiving sustenance; the eagles that nest in their enormous crowns when the transplants grow older; and many other species discussed in this book. While we can certainly transplant individual seedlings or even species, can we transplant whole ecosystems?

This white pine forest is only one example of the many ecosystems on earth where crucial decisions will soon need to be made that will decide their futures. Certainly, the success of our decisions will depend on our knowledge of the biology of each species in an ecosystem. Everything in conservation hinges in part on natural history. But the decisions we make are not purely ecological; they are societal, as well. These decisions will require us to respect all our stories, recognizing that they all have value and are vital to our success.

In the future, white pine will require our help to make the transition to a new global environment, as will many other species. To help white pine and other foundation species transition to a new environment, along with the ecosystems that depend on them, we will need people with the wisdom of Indigenous cultures as well as Henry David

Thoreau and George Perkins Marsh; the scientific acumen of Volney Spalding, Bernard Fernow, Gifford Pinchot, Perley Spaulding, and Bud Heinselman; the artistic vision of Sanford Gifford; and the practical know-how of the CCC, the Forest Service, sawmill owners like John Rajala, and conservation organizations like the Nature Conservancy. What we have learned about the natural and human history of white pine will help ecologists and conservationists face the challenges of the coming decades, serving as a guide to the future of white pine, other foundational species, and the cultures that spring from them.

Afterword

I have been living and working in the North Woods for most of my life and have come to love this beautiful forest more than any other place I have lived or visited. When you love a place this much, you want to tell people about it. I am an ecologist, so I like telling people about the natural history of the species that the North Woods comprise. Knowing the natural histories of these species enriches our experience of the North Woods, as it does for any other place.

My previous, and first, book with Island Press, *What Should a Clever Moose Eat?*, was a series of essays about the geological and natural history of the North Woods and the species in it. While I was writing it, I thought several times about including an essay on the natural history of white pine. But each time I started one, it seemed that a single essay would not do. White pine is connected to so many other species that it depends on and that in turn depend on it. It seemed that I would need several, perhaps many, essays to encompass the natural history of white pine and why it is a foundation of the North Woods ecosystem.

I also thought about how white pine became an important part of the culture and values of people who have lived and worked in the North

Woods. They include the Indigenous peoples, who revered white pine long before the arrival of European settlers; the lumberjacks with their heroic tales, which inspired the legend of Paul Bunyan; and the many generations of scientists and conservationists who worked hard to preserve what white pine remained and restore it to its former glory. The stories of these people contribute to what we feel when we approach a large, old white pine and, as people always do, put a hand on its trunk and look up into its crown with awe.

It gradually became clear that white pine needed its own book, one that would be an extended essay about how its natural history is intertwined with the cultures of humans who interact with it and how those cultures have important roles to play in its future in a rapidly changing world. If *What Should a Clever Moose Eat?* is a broad survey of the many species that dwell in the North Woods, then *White Pine* is a deeper exploration of one species that is a foundation of the North Woods ecosystem and the cultures that inhabit it. If you don't live in the North Woods, ask yourself which species play the same roles in ecology and culture where you live. And then ask yourself what can be done to care for them. As Robert Michael Pyle, a scientist, has said: "What we know we may choose to care for. What we fail to recognize, we certainly won't."[1]

This book and my thinking about white pine have benefited from the help of a number of people, and I would like to thank them here.

My wife, Mary Dragich, has read all the chapters and, with patience and fortitude, corrected me from many grammatical embarrassments. I seem to have a penchant for unclear antecedents and nonparallel tenses that she spots like a hawk. More important, she very diplomatically called attention to passages that didn't make sense to her and, therefore, might have been similarly confusing to you, the reader, were they not revised. Correcting the grammar and clarifying these passages helped sharpen my thinking about white pine throughout these pages. Her love and support during the two years that it took to write this book are priceless to me.

John Aber and I worked together in a series of old-growth forests when we were both at the University of Wisconsin, one of which was a magnificent white pine stand. We both still have fond memories of that time. When John retired from the University of New Hampshire a year ago, all his colleagues came together for a celebration on Zoom. (This was during the COVID-19 pandemic, after all.) At that celebration of John's career, I learned that he, too, was writing a book about climate change for what used to be called "the intelligent layperson," which is also you, the reader of this book. We decided to swap chapters as we wrote them and provide guidance and encouragement to each other. Working with John again was a pure joy for me and I hope also for him.

Meredith Cornett, once chief scientist and now climate change director for the Minnesota–North Dakota–South Dakota chapter of the Nature Conservancy, devoted an enjoyable afternoon to showing me the wonderful forests on her family's land in northern Minnesota, where there are large white pines as well as stumps of those that were logged. Our conversations on that hike inspired many thoughts on how climate change might affect white pine and the forests it inhabits—and what we can do about it. These ideas form the core of chapter 11. Meredith read that and several other chapters and noted valuable comments and references for them, and I thank her.

I have also benefited from and greatly enjoyed walking through the woods and talking about white pine and its management and restoration with John Rajala, of the fifth generation of the Rajala family. The Rajala family manages a beautiful chunk of the North Woods in Minnesota known as the Wolf Lake Tract, which contains many incredible white pines as well as sugar maple, red oak, yellow and paper birch, and other species. I drool over the lumber that comes out of John's sawmill, it is so gorgeous. It is great fun and a real privilege to be guided through these woods and through the Rajala family's local history, which started in the days of axes, oxen, and river drives. Chapter 10 of this book details these walks John and I have taken through the Wolf Lake Tract. The illustration for this chapter is taken partly from a historic photo from the Rajala family archives. What a treat is has been to hear about John's plans for taking his forest into the future. I hope chapter 10 does justice to the incredible work of John, his family, and his employees.

I also owe a great debt of thanks to the people at Island Press, especially Rebecca Bright, who edited the manuscript and gently pointed out places that need more attention and suggested ways to fix them. More important, Rebecca guided my thinking about the organization and scope of the book through several proposal iterations. The book has come a long way since we discussed the idea for it at the 2017 Annual Meeting of the Ecological Society of America. *White Pine* is much better for Rebecca's encouragement and helpful insights. As usual, the rest of the staff of Island Press also provided courteous professional help with the production of the book. Eliani Torres provided superb copyediting. Sharis Simonian guided the manuscript through the production stage with grace and patience. I am always amazed at how much work is involved at the production stage to create the handsome volume you now hold in your hands. All of you of the Island Press staff again deserve my highest gratitude.

Bibliography

Aber, John D., Jerry M. Melillo, Knute J. Nadelhoffer, Charles A. McClaugherty, and John Pastor. "Fine Root Turnover in Forest Ecosystems in Relation to Quantity and Forms of Nitrogen Availability: A Comparison of Two Methods." *Oecologia* 66, no. 3 (June 1985): 317–21. https://doi.org/bfcfh7.

Abrams, Marc D. "Eastern White Pine Versatility in the Presettlement Forest." *BioScience* 51, no. 11 (November 2001): 967–79. https://doi.org/fhn9br.

Albion, Robert Greenhalgh. *Forests and Sea Power: The Timber Problem of the Royal Navy 1652–1862.* Cambridge, MA: Harvard University Press, 1926.

Anderson, Mark G., Melissa Clark, and Arlene Olivero Sheldon. "Estimating Climate Resilience for Conservation across Geophysical Settings." *Conservation Biology* 28, no. 4 (August 2014): 959–70. https://doi.org/f6bhdj.

Asselin, Hugo. "Indigenous Forest Knowledge." Chap. 41 in *Routledge Handbook of Forest Ecology*, edited by Kelvin S.-H. Peh, Richard T. Corlett, and Yves Bergeron. London: Routledge, 2015.

Aubin, I., C. M. Garbe, S. Columbo, C. R. Drever, D. W. McKenney, C. Messier, J. Pedlar, et al. "Why We Disagree about Assisted Migration: Ethical Implications of a Key Debate regarding the Future of Canada's

Forests." *Forestry Chronicle* 87, no. 6 (December 2011): 755–65. https://doi.org/hrm3.

Barton, Andrew M., Alan S. White, and Charles V. Cogbill. *The Changing Nature of the Maine Woods*. Durham: University of New Hampshire Press, 2012.

Baskerville, Gordon L. "Redevelopment of a Degrading Forest System." *Ambio* 17, no. 5 (1988): 314–22. https://www.jstor.org/stable/4313487.

Benson, David R. *Stories in Log and Stone*. Saint Paul: Minnesota Department of Natural Resources, 2002.

Berkes, Fikret. *Sacred Ecology*. 2nd ed. London: Routledge, 2008.

Berkes, Fikret, and Iain J. Davidson-Hunt. "Biodiversity, Traditional Management Systems, and Cultural Landscapes: Examples from the Boreal Forests of Canada." *International Social Science Journal* 58, no. 187 (March 2006): 35–47. https://doi.org/d8fnfk.

Bonner, John Tyler. *Life Cycles: Reflections of an Evolutionary Biologist*. Princeton, NJ: Princeton University Press, 1993.

Brinkley, Douglas. *Rightful Heritage: Franklin D. Roosevelt and the Land of America*. New York: HarperCollins, 2016.

———. *Wilderness Warrior: Theodore Roosevelt and the Crusade for America*. New York: Harper Perennial, 2010.

Brown, Daniel. *Under a Flaming Sky: The Great Hinckley Firestorm of 1894*. Guilford, CT: Lyons, 2016.

Burns, Russell M., and Barbara H. Honkala, eds. *Silvics of North America*. Washington, DC: USDA Forest Service, 1990.

Canfield, Michael R. *Theodore Roosevelt in the Field*. Chicago: University of Chicago Press, 2015.

Carlson, Bradley Z., Jeffrey S. Munroe, and Bill Hegman. "Distribution of Alpine Tundra in the Adirondack Mountains of New York, U.S.A." *Arctic, Antarctic, and Alpine Research* 43, no. 3 (August 2011): 331–42. https://doi.org/c53nfc.

Chura, Patrick. *Thoreau the Land Surveyor*. Gainesville: University Press of Florida, 2010.

Cicero. *On the Republic. On the Laws*. Translated by Clinton W. Keyes. Loeb Classical Library 213. Cambridge, MA: Harvard University Press, 1928.

Clements, Frederic E. *Dynamics of Vegetation: Selections from the Writings of Frederic E. Clements*. Compiled by B. W. Allred and Edith S. Clements. New York: H. W. Wilson, 1949.

———. *The Life History of Lodgepole Burn Forests*. USDA Forest Service, Bulletin 79. Washington, DC: US Government Printing Office, 1910.

———. "Nature and Structure of the Climax." *Journal of Ecology* 24, no. 1 (February 1936): 252–84. https://doi.org/bvmft3.

Dahir, S. E., and J. E. Cummings Carlson. "Incidence of White Pine Blister Rust in a High-Hazard Region of Wisconsin." *Northern Journal of Applied Forestry* 18, no. 3 (June 2001): 81–86. https://doi.org/hrm5.

Davidson-Hunt, Iain J. "Indigenous Land Management, Cultural Landscapes and Anishinaabe People of Shoal Lake, Northwestern Ontario, Canada." *Environments* 31, no. 1 (January 2003): 21–42.

Davis, Margaret B. "Holocene Vegetation History of the Eastern United States." Chap. 11 in *Late-Quaternary Environments of the United States*. Vol. 2, edited by H. E. Wright Jr. Minneapolis: University of Minnesota Press, 1983.

——"Quaternary History and the Stability of Forest Communities." Chap. 10 in *Forest Succession: Concepts and Application*, edited by D. C. West, H. H. Shugart, and D. B. Botkin. Springer Advanced Texts in Life Sciences. New York: Springer-Verlag, 1981. https://doi.org/fj92xq.

Davis, Margaret B., and Ruth G. Shaw. "Range Shifts and Adaptive Responses to Quaternary Climate Change." *Science* 292 (April 27, 2001): 673–79. https://doi.org/c586g5.

Davis, Margaret B., Ruth G. Shaw, and Julie R. Etterson. "Evolutionary Responses to Climate Change." *Ecology* 86, no. 7 (July 2005): 1704–14. https://doi.org/cqm9ht.

Dean, Bradley P. "Henry D. Thoreau and Horace Greeley Exchange Letters on the 'Spontaneous Generation of Plants.'" *New England Quarterly* 66, no. 4 (December 1993): 630–38. https://doi.org/cbmczp.

Demuth, Bathsheba. *Floating Coast: An Environmental History of the Bering Strait*. New York: W. W. Norton, 2019.

Dewar, Roderick C. "Analytical Model of Carbon Storage in the Trees, Soils, and Wood Products of Managed Forests." *Tree Physiology* 8, no. 3 (April 1991): 239–58. https://doi.org/hrpx.

———. "A Model of Carbon Storage in Forests and Forest Products." *Tree Physiology* 6, no. 4 (December 1990): 417–28. https://doi.org/gdp84w.

Dobrowski, Solomon Z. "A Climatic Basis for Microrefugia: The Influence of Terrain on Climate." *Global Change Biology* 17, no. 2 (February 2011): 1022–35. https://doi.org/fbn2cm.

Dobson, Andrew. "Population Dynamics of Pathogens with Multiple Host Species." *American Naturalist* 164, no. S5 (November 2004): S64–S78. https://doi.org/fvk3hv.

Eckert, Andrew J., and Benjamin D. Hall. "Phylogeny, Historical Biogeography, and Patterns of Diversification for *Pinus* (Pinaceae): Phylogenetic Tests of Fossil Based Hypotheses." *Molecular Phylogenetics and Evolution* 40 (April 2006): 166–82. https://doi.org/dbzfcg.

Ellis, Christopher J., and D. Brian Deller. "Paleo-Indians." Chap. 3 in *The Archeology of Southern Ontario to A.D. 1650*, edited by Christopher J. Ellis and Neal Ferris. London: Ontario Archeological Society, London Chapter, 1990.

Ellison, Aaron M. "Foundation Species, Non-trophic Interactions, and the Value of Being Common." *iScience* 13 (March 29, 2019): 254–68. https://doi.org/ggbt9f.

Etterson, Julie R., Meredith W. Cornett, Mark A. White, and Laura C. Kavajecz. "Assisted Migration across Fixed Seed Zones Detects Adaptation Lags in Two Major North American Tree Species." *Ecological Applications* 30, no. 5 (July 2020): e02092. https://doi.org/gg2n7p.

Etterson, Julie R., and Ruth G. Shaw. "Constraint to Adaptive Evolution in Response to Global Warming." *Science* 294, no. 5540 (October 5, 2001): 151–54. https://www.jstor.org/stable/3084790.

Ewan, Nesta Dunn. "John Banister (1649 or 1650–1692)." *Encyclopedia Virginia*. Virginia Humanities. December 22, 2021. https://encyclopediavirginia.org/entries/banister-john-1649-or-1650-1692.

Fahey, Robert T., and Craig G. Lorimer. "Restoring a Midtolerant Pine Species as a Component of Late-Successional Forests: Results of Gap-Based Planting Trials." *Forest Ecology and Management* 292 (March 15, 2013): 139–49. https://doi.org/f4sbq7.

Falcon-Lang, Howard J., Viola Mages, and Margaret Collinson. "The Oldest *Pinus* and Its Preservation by Fire." *Geology* 44, no. 4 (April 2016): 303–6. https://doi.org/f8gtjh.

Fargione, Joseph E., Steven Bassett, Timothy Boucher, Scott D. Bridgham, Richard T. Conant, Susan C. Cook-Patton, Peter W. Ellis, et al. "Natural Climate Solutions for the United States." *Science Advances* 4, no. 11 (November 14, 2018). https://doi.org/gfnx72.

Foster, David, ed. *Hemlock: A Forest Giant on the Edge*. New Haven, CT: Yale University Press, 2014.

Frelich, Lee E., and Peter B. Reich. "Neighborhood Effects, Disturbance, and Succession in Forests of the Western Great Lakes Region1." Écoscience 2, no. 2 (1995): 148–58. https://doi.org/gf25fv.

Galatowitsch, Susan, Lee E. Frelich, and Laura Phillips-Mao. "Regional Climate Change Adaptation Strategies for Biodiversity Conservation in a Midcontinental Region of North America." *Biological Conservation* 142, no. 10 (October 2009): 2012–22. https://doi.org/d5xs2b.

Garret, Peter W., Ernst J. Schreiner, and Harry Kettlewood. *Geographic Variation of White Pine in the Northeast*. US Forest Service Research Paper NE-274. Upper Darby, PA: Northeast Forest Experiment Station, 1973.

Geils, Brian W., Kim E. Hummer, and Richard S. Hunt. "White Pines, *Ribes*, and Blister Rust: A Review and Synthesis." *Forest Pathology* 40, no. 3–4 (August 2010): 147–85. https://doi.org/csmbnw.

Genys, John B. "Geographic Variation in Eastern White Pine: Two-Year Results of Testing Wide-Range Collections in Maryland." *Silvae Genetica* 17, no. 1 (1968): 6–12. https://www.jstor.org/stable/4033120.

Gernandt, David S., Gretel G. López, Sol O. Garcia, and Aaron Liston. "Phylogeny and Classification of *Pinus*." *Taxon* 54, no. 1 (February 2005): 29–42. https://doi.org/bncm3z.

Gevorkiantz, S. R., and Raphael Zon. *Second-Growth White Pine in Wisconsin: Its Growth, Yield, and Commercial Possibilities*. Research Bulletin 98. Madison: Agricultural Experiment Station of the University of Wisconsin, Lakes States Forest Experiment Station of the Forest Service (USDA), Wisconsin Conservation Commission, 1930.

Gray, Asa. *A Manual of the Botany of the Northern United States*. Boston: James Munroe, 1848.

Grinnell, Joseph. *Joseph Grinnell's Philosophy of Nature*. Berkeley: University of California Press, 1943.

Hacskaylo, Edward, and Albert G. Snow Jr. *Relation of Soil Nutrients and Light to Prevalence of Mycorrhizae on Pine Seedlings.* Station Paper NE-125. Upper Darby, PA: USDA Forest Service, Northeast Forest Experiment Station, 1959.

Hämäläinen, Pekka. *Lakota America: A New History of Indigenous Power.* The Lamar Series in Western History. New Haven, CT: Yale University Press, 2019.

Harmon, Mark E., William K. Ferrell, and Jerry F. Franklin. "Effects on Carbon Storage of Conversion of Old-Growth Forests to Young Forests." *Science* 247, no. 4943 (February 9, 1990): 699–702. https://doi.org/ffg95d.

Harper, John L. *Population Biology of Plants.* New York: Academic Press, 1977.

He, Tianhua, Juli G. Pausus, Claire M. Belcher, Dylan W. Schwilk, and Byron B. Lamont. "Fire-Adapted Traits of *Pinus* Arose in the Fiery Cretaceous." *New Phytologist* 194, no. 3 (May 2012): 751–59. https://doi.org/gg976v.

Heinselman, Miron L. *The Boundary Waters Wilderness Ecosystem.* Minneapolis: University of Minnesota Press, 1996.

———. "Fire in the Virgin Forests of the Boundary Waters Canoe Area, Minnesota." *Quaternary Research* 3, no. 3 (1973): 329–82. https://doi.org/brhmvt.

———. "Fire Intensity and Frequency as Factors in the Distribution and Structure of Northern Ecosystems." In *Fire Regimes and Ecosystem Properties: Proceedings of the Conference Held December 11–15, 1978, Honolulu, Hawaii: General Technical Report WO26*, edited by H. A. Mooney et al., 7–57. Washington, DC: USDA Forest Service, 1981. https://www.researchgate.net/publication/266202894_Fire_in_tropical_ecosystems.

———. "Forest Sites, Bog Processes, and Peatland Types in the Glacial Lake Agassiz Region of Minnesota." *Ecological Monographs* 33, no. 4 (Autumn 1963): 327–74. https://doi.org/d9cczc.

———. "Landscape Evolution, Peatland Types, and the Environment in the Lake Agassiz Peatlands Natural Area, Minnesota." *Ecological Monographs* 40, no. 2 (Spring 1970): 235–61. https://doi.org/cg4cpw.

Heinselmann, Miron L., Lewis F. Ohmann, Robert R. Ream, and Charles T. Brown. "A Problem Analysis of Research in the Ecology of Wilderness

Biotic Communities in the Boundary Waters Canoe Area." Unpublished Report. April 1967. Miron L. Heinselman Papers, Minnesota Historical Society. Location 141.C.7.5B, box 2, folder "Origin of Ecological Research Program—NCFES 1964–1972."

Helvey, J. D. "Interception by Eastern White Pine." *Water Resources Research* 3, no. 3 (1967): 723–29. https://doi.org/c4xv27.

Hemingway, Ernest. *The Nick Adams Stories*. New York: Scribner, 1972.

Hoegh-Guldberg, O., L. Hughes, S. McIntyre, D. B. Lindemeyer, C. Parmesan, H. P. Possingham, and C. D. Thomas. "Assisted Colonization and Rapid Climate Change." *Science* 321, no. 5877 (July 18, 2008): 345–46. https://doi.org/fqmr35.

Holbrook, Stuart. *Holy Old Mackinaw: A Natural History of the American Lumberjack*. New York: Macmillan, 1956.

Horsfall, James G. "Roland Thaxter." *Annual Review of Phytopathology* 19 (September 1979): 29–35. https://doi.org/bjnmn2.

Hotchkiss, George W. *History of the Lumber and Forest Industry of the Northwest*. Chicago: G. W. Hotchkiss, 1898.

Huber, J. Parker. *The Wildest Country: Exploring Thoreau's Maine*. 2nd ed. With a foreword by Bill McKibben. Boston: Appalachian Mountain Club, 2008.

Hunter, Malcolm L. "Climate Change and Moving Species: Furthering the Debate on Assisted Colonization." *Conservation Biology* 21, no. 5 (October 2007): 1356–58. https://doi.org/ctvv6j.

Irvine, Karen H. "When the Women Manned the Mountain." *Northern Woodlands*, Summer 2021.

Iverson, Louis R., and Anantha Prasad. "Predicting Abundance of 80 Tree Species Following Climate Change in the Eastern United States." *Ecological Monographs* 68, no. 4 (1998): 465–85. https://doi.org/cr5grc.

Jacoby, Karl. "Class and Environmental History: Lessons from 'The War in the Adirondacks.'" *Environmental History* 2, no. 3 (July 1997): 324–42. https://doi.org/dx4q8r.

Johnson, Christopher, and David Govatski. *Forests for the People*. Washington, DC: Island Press, 2013.

Johnson, Kirk. "From a Woodland Elegy, a Rhapsody in Green; Hunter Mountain Paintings Spurred Recovery." *New York Times*, June 7, 2001.

https://www.nytimes.com/2001/06/07/nyregion/woodland-elegy-rhapsody-green-hunter-mountain-paintings-spurred-recovery.html.

Johnson, Kurt, and Steve Coates. *Nabokov's Blues*. Cambridge, MA: Noland Books, 1999.

Kalm, Peter. *Peter Kalm's Travels in North America: The English Version of 1770.* New York: Wilson-Erickson, 1937. Revised and edited by Adolph B. Benson, with a translation of new material from Kalm's diary notes. 2 vols. New York: Dover, 1966.

Kauffman, Erle. "Roosevelt—Forest Camp No. 1." *American Forests* 39, no. 6 (June 1933): 251–54.

Keller, Jane E. *Adirondack Wilderness: A Story of Man and Nature*. Syracuse, NY: Syracuse University Press, 1980.

Kershaw, John A., Mark J. Ducey, Thomas W. Beers, and Bertram Husch. *Forest Mensuration*. 5th edition. New York: Wiley-Blackwell, 2016.

King, D. B. *Incidence of White Pine Blister Rust in the Lake States*. Lakes States Forest Experiment Station Paper No. 64. USDA Forest Service. Washington, DC: US Government Printing Office, 1958.

Klironomos, John N., and Miranda M. Hart. "Animal Nitrogen Swap for Plant Carbon." *Nature* 410, no. 6829 (April 2001): 651–52. https://doi.org/bjsbw6.

Larson, Agnes M. *The White Pine Industry in Minnesota: A History*. 1949. Felser-Lampert Minnesota Heritage Book. Minneapolis: University of Minnesota Press, 2007.

Leopold, Aldo. "Wilderness as a Land Laboratory." In *The River of the Mother of God and Other Essays by Aldo Leopold*, edited by Susan L. Flader and J. Baird Callicott, 287–89. Madison: University of Wisconsin Press, 1991.

Linnaeus, Carl. *Species Plantarum, A Facsimile of the First Edition*. Vol. 2. London: Ray Society, 1959.

Liston, Aaron, William A. Robinson, Daniel Piñero, and Elena R. Alvarez-Bullya. "Phylogenetics of *Pinus* (Pinaceae) Based on Nuclear Ribosomal DNA Internal Transcribed Spacer Region Sequences." *Molecular Phylogenetics and Evolution* 11, no. 1 (February 1999): 95–109. https://doi.org/fsqdj5.

Lopez, Barry H. *Arctic Dreams*. New York: Charles Scribner's Sons, 1986.

Loucks, Orie L. "Evolution of Diversity, Efficiency, and Community Stability." *American Zoologist* 10, no. 1 (February 1970): 17–25. https://doi.org/c9t56j.

Maher, Neil M. *Nature's New Deal.* New York: Oxford University Press, 2008.

———. "A New Deal Body Politic: Landscape, Labor, and the Civilian Conservation Corps." *Environmental History* 7 (2002): 435–61. https://doi.org/cs9t8z.

Maloy, Otis C. "White Pine Blister Rust Control in North America: A Case History." *Annual Review of Phytopathology* 35 (1997): 87–109. https://doi.org/bbrg45.

Marsh, George P. *Man and Nature.* Seattle: University of Washington Press, 1965.

———. *Report on Artificial Propagation of Fish.* George Perkins Marsh Online Research Center. University of Vermont Libraries Digital Collections. https://cdi.uvm.edu/manuscript/uvmcdi-85595.

Marvel, Kate. "The Parallel Universes of a Woman in Science." *Nautilus* 41 (October 2016). https://nautil.us/the-parallel-universes-of-a-woman-in-science-5248/.

Maser, Chris, James M. Trappe, and Ronald A. Nussbaum. "Fungal-Small Mammal Interrelationships with Emphasis on Oregon Coniferous Forests." *Ecology* 59, no. 4 (Summer 1978): 799–809. https://doi.org/bg3f46.

McClaugherty, Charles A., John Pastor, John D. Aber, and Jerry M. Melillo. "Forest Litter Decomposition in Relationship to Soil Nitrogen Dynamics and Litter Quality." *Ecology* 66, no. 1 (February 1985): 266–75. https://doi.org/dq792f.

McClurken, James M., Charles E. Cleland, Thomas Lund, John D. Nichols, Helen Tanner, and Bruce White. *Fish in the Lakes, Wild Rice, and Game in Abundance: Testimony on Behalf of Mille Lacs Ojibway Hunting and Fishing Rights.* East Lansing: Michigan State University Press, 2000.

McEntee, James J. *Final Report of the Director of the Civilian Conservation Corps, April 1933 through June 30, 1942.* Washington, DC: National Archives and Records Administration, 1942.

McLachlan, Jason S., Jessica J. Hellman, and Mark W. Schwartz. "A

Framework for Debate of Assisted Migration in an Era of Climate Change." *Conservation Biology* 21, no. 2 (April 2007): 297–302. https://doi.org/dd2v62.

McPhee, John. *Rising from the Plains*. New York: Farrar, Straus and Giroux, 1987.

Meldahl, Keith H. *Rough-Hewn Land: A Geologic Journey from California to the Rocky Mountains*. Berkeley: University of California Press, 2013.

Millar, Constance I. "Early Evolution of Pines." Chap. 3 in *Ecology and Biogeography of* Pinus, edited by David M. Richardson. Cambridge: Cambridge University Press, 1998.

———. "Impact of the Eocene on the Evolution of *Pinus* L." *Annals of the Missouri Botanical Garden* 80, no. 2 (1993): 471–98. https://doi.org/btw6mh.

Millar, Constance I., Nathan L. Stephenson, and Scott L. Stephens. "Climate Change and Forests of the Future: Managing in the Face of Uncertainty." *Ecological Applications* 17, no. 8 (December 2007): 2145–51. https://doi.org/fdc4hb.

Miller, Char. *Gifford Pinchot and the Making of Modern Environmentalism*. Washington, DC: Island Press, 2001.

Minor, Jesse, and Geoffrey A. Boyce. "Smokey Bear and the Pyropolitics of United States Forest Governance." *Political Geography* 62 (January 2018): 79–93. https://doi.org/gc22gz.

Mitchell, H. L. "The Growth and Nutrition of White Pine (*Pinus strobus* L.) Seedlings in Cultures with Varying Nitrogen, Phosphorus, Potassium and Calcium." *Black Rock Forest, Bulletin No. 9*. Cornwall, NY: Cornwall Press, 1939.

Mitchell, H. L., R. F. Finn, and R. O. Rosendahl. "The Relation between Mycorrhizae and the Growth and Nutrient Absorption of Coniferous Seedlings in Nursery Beds." *Black Rock Forest Papers* 1, no. 10 (1937): 58–73.

Morison, Samuel E. *The Maritime History of Massachusetts*. Northeastern Classics Edition. Boston: Northeastern University Press, 1979. First published 1921 by Houghton Mifflin (Boston).

Nowak, Lisa. *Cultural Landscape Report for Eleanor Roosevelt National Historic Site*. Boston: Olmsted Center for Landscape Preservation, 2005.

Nuttle, Tim, Todd E. Ristau, and Alejandro A. Royo. "Long-Term Biological Legacies of Herbivore Density in a Landscape-Scale Experiment: Forest Understoreys Reflect Past Deer Density Treatments for at Least 20 Years." *Journal of Ecology* 102, no. 1 (January 2014): 221–28. https://doi.org/gpjtpf.

Oreskes, Naomi, ed. *Plate Tectonics: An Insider's History of the Modern Theory of the Earth*. Boca Raton, FL: CRC, 2003.

Overpeck, Jonathan T., Patrick J. Bartlein, and Thomas Webb III. "Potential Magnitude of Future Vegetation Change in Eastern North America: Comparisons with the Past." *Science* 254, no. 5032 (November 1, 1991): 692–95. https://doi.org/b83fdq.

Palik, Brian, Anthony W. D'Amato, Jerry F. Franklin, and K. Norman Johnson. *Ecological Silviculture: Foundations and Applications*. Long Grove, IL: Waveland, 2021.

Parker, Arthur C. "Certain Iroquois Tree Myths and Symbols." *American Anthropologist* 14, no. 4 (October–December 1912): 608–20. https://www.jstor.org/stable/pdf/659833.pdf.

Pastor, John. *What Should a Clever Moose Eat? Natural History, Ecology, and the North Woods*. Washington, DC: Island, 2016.

Pastor, John, John D. Aber, Charles A. McClaugherty, and Jerry M. Melillo. "Aboveground Production and N and P Cycling along a Nitrogen Mineralization Gradient on Blackhawk Island, Wisconsin." *Ecology* 65, no. 1 (February 1984): 256–68. https://doi.org/fngqww.

Pastor, John, Bradley Dewey, and Donald P. Christian. "Carbon and Nutrient Mineralization and Fungal Spore Composition of Fecal Pellets from Voles in Minnesota." *Ecography* 19, no. 1 (March 1996): 52–61. https://doi.org/c43bwb.

Pastor, John, and Wilfred M. Post. "Response of Northern Forests to CO_2-Induced Climatic Change." *Nature* 334 (July 7, 1988): 55–58. https://doi.org/ct2drb.

Peattie, Donald C. *A Natural History of Trees of Eastern and Central North America*. New York: Houghton-Mifflin, 1991.

Peckham, Howard H. *The Making of the University of Michigan*. Ann Arbor: University of Michigan Press, 1994.

Pedlar, John H., Daniel W. McKenny, Isabelle Aubin, Tannis Beardmore,

Jean Beaulieu, Louis Iverson, Gregory A. O'Neill, et al. "Placing Forestry in the Assisted Migration Debate." *BioScience* 62, no. 9 (September 2012): 835–42. https://doi.org/f39rtc.

Perry, D. A., R. Molina, and M. P. Amaranthus. "Mycorrhizae, Mycorrhizospheres, and Reforestation: Current Knowledge and Research Needs." *Canadian Journal of Forest Research* 17, no. 8 (August 1987): 929–40. https://doi.org/cjw4vj.

Pinchot, Gifford. *Breaking New Ground*. New York: Harcourt, Brace, 1947.

———. *Fishing Talk*. Harrisburg, PA: Stackpole Books, 1993.

Pinchot, Gifford, and Henry S. Graves. *The White Pine: A Study, with Tables of Volume and Yield*. New York: Century, 1896. https://hdl.handle.net/2027/hvd.32044026393561.

Plukenet, Leonard. *Almagestum Botanicum*. London: printed by the author, 1696. https://bibdigital.rjb.csic.es/records/item/10874-almagestum-botanicum.

Porter, Jess. "African Americans in the CCC." Civilian Conservation Corps in Arkansas. 2019. https://ualrexhibits.org/ccc/african-americans-in-the-ccc/.

Proescholdt, Kevin. "First Fight: Bud Heinselman and the Boundary Waters Canoe Area, 1964–1965." *Minnesota History* 64, no. 2 (Summer 2014): 70–84. https://collections.mnhs.org/mnhistorymagazine/articles/64/v64i02p070-084.pdf.

Proescholdt, Kevin, Rip Rapson, and Miron L. Heinselman. *Troubled Waters*. Saint Cloud, MN: North Star, 1996.

Puettmann, Klaus J., K. David Coates, and Christian Messier. *A Critique of Silviculture: Managing for Complexity*. Washington, DC: Island Press, 2009.

Pyle, Robert Michael. "The Rise and Fall of Natural History." In *The Future of Nature: Writing on a Human Ecology from Orion Magazine*, edited by Barry H. Lopez, 233–43. Minneapolis: Milkweed Editions, 2007.

Rajala, Benhart. *Tim-BERRR! Pine Logging in the Big Fork River Country*. 2 vols. Saint Cloud, MN: North Star, 1992.

Rajala, Jack. *Bringing Back the White Pine*. Grand Rapids, MN: self-published, 1998.

Rajala, John. "Bud Capping Sept 4 2016." September 4, 2016. YouTube video, 3:05. https://youtu.be/coG4alpnqDg.

———. "Rajala Forestry—Northern Red Oak Shelterwood." September 10, 2016. YouTube video, 23:01. https://youtu.be/VQAgjRbCdA8.

———. "Why We Bud Cap White Pine at Rajala Forestry." *Northwoods Notebook* (blog). Minnesota Timber & Millwork, Rajala Forestry. January 3, 2021. https://mntimber.com/blog/f/why-we-bud-cap-white-pine-at-rajala-forestry.

Reed, Howard S. "Volney Morgan Spalding." *Plant World* 22, no. 1 (January 1919): 14–18. https://www.jstor.org/stable/43477698.

Reiners, William A., and Kenneth L. Dreise. *Transport Processes in Nature: Propagation of Ecological Influences through Environmental Space*. Cambridge: Cambridge University Press, 2004.

Ricciardi, Anthony, and Daniel Simberloff. "Assisted Colonization Is Not a Viable Conservation Strategy." *Trends in Ecology & Evolution* 24, no. 5 (May 2009): 248–53. https://doi.org/czfhpk.

Rice, Ronald E. "Smokey Bear." Chap. 17 in *Public Communication Campaigns*. 3rd ed. Edited by Ronald E. Rice and Charles K. Atkins. Thousand Oaks, CA: SAGE, 2001.

Richter, Daniel K. *The Ordeal of the Longhouse: The Peoples of the Iroquois League in the Era of European Colonization*. Chapel Hill: Omohundro Institute and the University of North Carolina Press, 1992.

Ritchie, William A., and Robert E. Funk. "Paleo-Indians in a New Perspective: Comments on the Assembled Papers." *Archaeology of Eastern North America* 12 (1984): 1–4. https://www.jstor.org/stable/40914230.

Robbins, Kathryn, William A. Jackson, and Ronald E. McRoberts. "White Pine Blister Rust in the Eastern Upper Peninsula of Michigan." *Northern Journal of Applied Forestry* 5, no. 4 (December 1988): 263–64. https://doi.org/hrqh.

Roberts, Kenneth. *Arundel*. Garden City, NY: Doubleday, 1930.

———. *Trending into Maine*. Garden City, NY: Doubleday, 1938.

Rodgers, Andrew D. *Bernhard Eduard Fernow: A Story of North American Forestry*. Durham, NC: Forest History Society, 1991.

Rossier, James. "A True Relation of Captain George Weymouth, His Voyage Made This Present Yeare 1605; in the Discouerie of the North Part of Virginia." In *A History of Maine: A Collection of Readings on the History of*

Maine 1600–1976, 4th ed., edited by Ronald F. Banks, 7–22. Dubuque, IA: Kendall/Hunt, 1976.

Russell, Emily W. B. "Indian-Set Fires in Forests of the Northeastern United States." *Ecology* 64, no. 1 (February 1983): 78–88. https://doi.org/cfjkhr.

Rutkow, Eric. *American Canopy: Trees, Forests, and the Making of a Nation.* New York: Scribner, 2012.

Ryan, Michael G., Dan Binkley, and James H. Fownes. "Age-Related Decline in Forest Productivity: Pattern and Process." *Advances in Ecological Research* 27 (1997): 213–62. https://doi.org/djt7rs.

Saladin, Bianca, Andrew B. Leslie, Rafael O. Wüest, Glenn Litsios, Elena Conti, Nicolas Salamin, and Niklaus E. Zimmerman. "Fossils Matter: Improved Estimates of Divergence Times in *Pinus* Reveal Older Diversification." *Evolutionary Biology* 17 (2017): 95–110. https://doi.org/f93d48.

Samuelson, Paul A. "The Economics of Forestry in an Evolving Society." *Economic Inquiry* 14, no. 4 (December 1976): 466–92. https://doi.org/hrqj.

Seltzer, Alan M., Jessica Ng, Werner Aeschbach, Rolf Kipfer, Justin T. Kulongowski, Jeffrey P. Severinghaus, and Martin Stute. "Widespread Six Degrees Celsius Cooling on Land during the Last Glacial Maximum." *Nature* 593 (May 12, 2021): 228–32. https://doi.org/gjz4wz.

Sharma, Mahadev, and John Parton. "Modelling the Effects of Climate on Site Productivity of White Pine Plantations." *Canadian Journal of Forest Research* 49, no. 10 (2019): 1289–97. https://doi.org/hrqk.

Simard, Suzanne. *Finding the Mother Tree: Discovering the Wisdom of the Forest.* New York: Alfred A. Knopf, 2021.

The Sloane Herbarium: An Annotated List of the Horti Sicci *Composing It; With Biographical Accounts of the Principal Contributors.* Edited by J. E. Dandy from materials compiled by James Britten. With an introduction by Spencer Savage. London: British Museum (Natural History), 1958.

Spalding, Volney M. "The Rise and Progress of Ecology." *Science* 17, no. 423 (February 6, 1903): 201–10. https://doi.org/fk35cw.

Spalding, Volney M., and Bernhard E. Fernow. *The White Pine.* Washington, DC: US Government Printing Office, 1899. https://www.biodiversitylibrary.org/bibliography/17156#/summary.

Spaulding, Perley. *The Blister Rust of White Pine*. US Department of Agriculture Bulletin No. 206. Washington, DC: US Government Printing Office, 1911.

———. *Investigations of the White-Pine Blister Rust*. US Department of Agriculture Bulletin No. 957. Washington, DC: US Government Printing Office, 1922.

Stea, Rudolph R., and Susan E. Pullan. "Hidden Cretaceous Basins in Nova Scotia." *Canadian Journal of Earth Science* 38, no. 9 (September 2001): 1335–54. https://doi.org/dmv7jp.

Stegner, Wallace. *Where the Bluebird Sings to the Lemonade Springs*. New York: Random House, 1992.

Swank, Wayne T., and N. H. Miner. "Conversion of Hardwood-Covered Watersheds to White Pine Reduces Water Yield." *Water Resources Research* 4, no. 5 (October 1968): 947–54. https://doi.org/dds69m.

Swift, Lloyd W., Jr., Wayne T. Swank, J. B. Mankin, R. J. Luxmoore, and R. A. Goldstein. "Simulation of Evapotranspiration and Drainage from Nature and Clear-Cut Deciduous Forests and Young Pine Plantation." *Water Resources Research* 11, no. 5 (October 1975): 667–73. https://doi.org/cn2hc7.

Terwilliger, John, and John Pastor. "Small Mammals, Ectomycorrhizae, and Conifer Succession in Beaver Meadows." *Oikos* 85, no. 1 (April 1999): 83–94. https://doi.org/cv3272.

Tester, John, Anthony M. Starfield, and Lee E. Frelich. "Modeling for Ecosystem Management in Minnesota Pine Forests." *Biological Conservation* 80, no. 3 (June 1997): 313–24. https://doi.org/d88x8g.

Thiers, Barbara M. *Herbarium*. Portland, OR: Timber, 2020.

Thoreau, Henry David. *Faith in a Seed: The Dispersion of Seeds and Other Late Natural History Writing*. Edited by Bradley P. Dean. Washington, DC: Island Press, 1993.

———. *The Journal of Henry D. Thoreau*. New York: Houghton Mifflin, 1906. Facsimile of the first edition, edited by Bradford Torrey and Francis H. Allen. 14 vols. bound as 2 vols. (1–7, 8–14). New York: Dover, 1962.

———. *The Maine Woods: A Fully Annotated Edition*. Edited by Jeffrey S. Cramer. New Haven, CT: Yale University Press, 2009.

———. *The Natural History Essays*. Literature of the American Wilderness. Edited by Robert Sattelmeyer. Salt Lake City, UT: Peregrine Smith Books, 1980.

———. *Walden: A Fully Annotated Edition*. Edited by Jeffrey S. Cramer. New Haven, CT: Yale University Press, 2004.

Trombulak, Stephen C. *So Great a Vision: The Conservation Writings of George Perkins Marsh*. Hanover, NH: Middlebury College Press, 2001.

Trosper, Ronald L. "Indigenous Influence on Forest Management on the Menominee Indian Reservation." *Forest Ecology and Management* 249, no. 1–2 (September 2007): 134–39. https://doi.org/cvdsww.

Trow, George W. S. *The Harvard Black Rock Forest*. Sightline Books: The Iowa Series in Literary Nonfiction. Iowa City: University of Iowa Press, 2004.

USDA Forest Service. *Fire Management Action Plan*. Duluth, MN: Superior National Forest, 1991.

———. "Tree Atlas, version 4." Climate Change Atlas. Northern Research Station. https://www.fs.fed.us/nrs/atlas/tree/.

Uprety, Yadav, Hugo Asselin, and Yves Bergeron. "Cultural Importance of White Pine (*Pinus strobus*) to the Kitcisakik Algonquin Community of Western Quebec, Canada." *Canadian Journal of Forest Research* 43, no. 6 (June 2013): 544–51. https://doi.org/f42vc2.

Van Arsdel, E. P. "The Nocturnal Diffusion and Transport of Spores." *Phytopathology* 57, no. 11 (January 1967): 1221–29.

Vietze, Andrew. *White Pine: American History and the Tree That Made a Nation*. Guilford, CT: Globe Pequot, 2018.

Walls, Laura Dassow. "'As Planets Faithful Be': The Higher Law of Science in Emerson's Antislavery Lectures." *Nineteenth-Century Prose* 30, nos. 1/2 (Spring/Fall 2003): 171–94.

———. *Henry David Thoreau: A Life*. Chicago: University of Chicago Press, 2017.

———. *Seeing New Worlds: Henry David Thoreau and Nineteenth-Century Natural Science*. Madison: University of Wisconsin Press, 1995.

Wein, Ross W., and Mohammed A. El-Bayoumi. "Limitations to Predictability of Plant Succession in Northern Ecosystems." In *Resources and Dynamics of the Boreal Zone Proceedings of a Conference Held at Thunder*

Bay, Ontario, August 1982, edited by Ross W. Wein, Roderick R. Riewe, and Ian R. Methven, 214–25. Ottawa, ON: Association of Canadian Universities for Northern Studies, 1983.

Wendel, G. W., and H. C. Smith. "*Pinus strobus* L. Eastern White Pine." In *Conifers*, edited by Russell M. Burns and Barbara H. Honkala, 476–88. Vol. 1 of *Silvics of North America*. Washington, DC: USDA Forest Service, 1990.

Weyenberg, Scott A., Lee E. Frelich, and Peter B. Reich. "Logging versus Fire: How Does Disturbance Type Influence the Abundance of *Pinus strobus* Regeneration?" *Silva Fennica* 38, no. 2 (2004): 179–94. https://doi.org/hrqm.

White, Mark A. "Long-Term Effects of Deer Browsing: Composition, Structure and Productivity in a Northeastern Minnesota Old-Growth Forest." *Forest Ecology and Management* 269 (2012): 222–28. https://doi.org/ggz68t.

White, Mark A., Meredith W. Cornett, Katie Frerker, and Julie R. Etterson. "Partnerships to Take on Climate Change: Adaptation Forestry and Conifer Strongholds Projects in the Northwoods, Minnesota, USA." *Journal of Forestry* 118, no. 3 (May 2020): 219–32. https://doi.org/gm662n.

Williamson, James A. *The Voyages of the Cabots and the English Discovery of North America under Henry VII and Henry VIII*. London: Argonaut, 1929.

Winder, Richard S., Elizabeth A. Nelson, and Tannis Beardmore. "Ecological Implications for Assisted Migration in Canadian Forests." *Forestry Chronicle* 87, no. 6 (December 2011): 731–44. https://doi.org/hrqn.

Zinck, John W. R., and Om P. Rajora. "Post-Glacial Phylogeography and Evolution of a Wide-Ranging Highly-Exploited Keystone Forest Tree, Eastern White Pine (*Pinus strobus*) in North America: Single Refugium, Multiple Routes." *BMC Evolutionary Biology* 16 (March 2016): 56–73. https://doi.org/f8bwnk.

Notes

Introduction

1. Pastor et al., "N and P Cycling," 256–68; McClaugherty et al., "Forest Litter Decomposition," 266–75; Aber et al., "Fine Root Turnover," 317–21.
2. For more details about sizes and environmental preferences of white pine, see Wendel and Smith, "*Pinus strobus* L.," 476–88; Sharma and Parton, "Site Productivity," 1289–97; Abrams, "Eastern White Pine Versatility," 967–79; USDA Forest Service, "Tree Atlas, version 4."
3. For an analysis of hemlock as a foundation species in the North Woods, see Foster, *Hemlock*.
4. For a review of the idea of foundation species, see Ellison, "Foundation Species, Non-trophic Interactions," 254–68.
5. Peattie, *Natural History of Trees*, 3.

Chapter 1. The Evolution and Arrival of White Pine

1. Marvel, "Parallel Universes."
2. For an excellent introduction to the history of the theory of plate tectonics, see Oreskes, *Plate Tectonics*.
3. Falcon-Lang, Mages, and Collinson, "Preservation by Fire," 303–6.
4. Except for one species (*Pinus merkusii*) found in the highlands of Burma

and Sumatra, which probably spread south from an ancestor in far eastern Eurasia.

5. Stea and Pullan, "Hidden Cretaceous Basins," 1335–54.
6. He et al., "Fire-Adapted Traits of *Pinus*," 751–59.
7. *Pinus* and *Strobus* are further subdivided into sections and subsections that in turn encompass all the pine species. Ideally, the classification of pines into these groupings reflects the evolutionary descent of modern pine species from the last common ancestor, at this time thought to be *Pinus mundayi*.

   Classical taxonomy going back to Carl Linnaeus based these classifications on visible traits, especially those of flowers and cones, but with the rise of inexpensive and efficient molecular methods, modern classifications are based more on DNA sequences. For the development of DNA-based classifications of pines and their evolution, see especially Liston et al., "Phylogenetics of *Pinus* (Pinaceae)," 95–109; Gernandt et al., "Classification of *Pinus*," 29–42; Eckert and Hall, "Patterns of Diversification," 166–82; and Saladin et al., "Fossils Matter," 95–110.
8. Millar, "Evolution of *Pinus* L.," 471–98; Millar, "Early Evolution of Pines," chap. 3.
9. Eckert and Hall, "Phylogeny, Historical Biogeography," 166–82.
10. Exactly how or where the Rocky Mountains served as refugia for pine species is not so simple as this sentence implies, although the broad outline worked out by Millar ("Evolution of *Pinus* L.," 471–98; "Early Evolution of Pines," chap. 3) is generally accepted. The Rocky Mountains are not a single mountain chain but a complicated cordillera of more than a dozen mountain ranges, many of which are separated by downfaulted and then partly filled basins while others (e.g., the Absaroka Range) overlie partly buried ranges (e.g., the Washakie Range). The full geological history of this complex mountainous terrain, as well as the details of how it guided pine speciation, still has not been completely worked out. For readable accounts of how plate tectonics built the Rockies, see McPhee, *Rising from the Plains*, and Meldahl, *Rough-Hewn Land*.
11. Millar, "Evolution of *Pinus* L.," 471–98; Millar, "Early Evolution of Pines," chap. 3.
12. Millar, "Evolution of *Pinus* L.," 471–98; Millar, "Early Evolution of Pines," chap. 3.
13. *Journal of Henry D. Thoreau*, "June 21, 1860," 1643.

14. Davis, "Stability of Forest Communities," 132–53; Davis, "Holocene Vegetation History," chap. 11.
15. Zinck and Rajora, "Post-Glacial Phylogeography and Evolution," 56–73.
16. Figure redrawn and simplified from Davis, "Holocene Vegetation History," figure 11.6; Zinck and Rajora, "Post-Glacial Phylogeny and Evolution," figure 6.
17. Genys, "Geographic Variation," 6–12; Garret, Schreiner, and Kettlewood, *Variation of White Pine*, 1–4.

Chapter 2. "A Great Store of Wood and Above All of Pines"

1. Ellis and Deller, "Paleo-Indians," chap. 3. See also Ritchie and Funk, "A New Perspective," and papers therein.
2. Later, a sixth nation, the Tuscarora, joined the Iroquois Confederacy. For a history of the Iroquois Confederacy and its dealings with European settlers, see Richter, *Ordeal of the Longhouse*.
3. Parker, "Certain Iroquois Tree Myths," 608–20.
4. Uprety, Asselin, and Bergeron, "Cultural Importance of White Pine," 544–51; Asselin, "Indigenous Forest Knowledge," chap. 41.
5. Berkes, *Sacred Ecology*, 7.
6. Williamson, *Voyages of the Cabots*, 202–3.
7. Williamson, 41.
8. Rossier, "Captain George Weymouth," 18.
9. Plukenet, *Almagestum Botanicum*, 296.
10. Thiers, *Herbarium*.
11. Linnaeus, *Species Plantarum*, vol. 2, 1001.
12. Ewan, "John Banister."
13. *Sloane Herbarium*, 183.
14. Albion, *Forests and Sea Power*, 28.
15. Albion, 233.
16. Albion, 29.
17. Vietze, *White Pine*, 65.
18. Albion, *Forests and Sea Power*, 248.
19. Albion, 249.

20. Albion, 282.

Chapter 3. A Logger's Paradise

1. For an enjoyable and generally historically accurate novel of this campaign, including excellent descriptions of the North Woods of the Penobscot and Kennebec River valleys at this time, see Roberts, *Arundel.*
2. A board foot is the standard measure of lumber and is a board one foot wide, one foot long, and one inch thick. A cord is the standard measure of smaller logs to be ground for pulp or split for firewood and is a stack of wood four feet wide, four feet high, and eight feet long. Wendel and Smith, "*Pinus strobus* L.," 476–78.
3. Hämäläinen, *Lakota America.*
4. For a detailed analysis of treaties between the Ojibwa and the State of Minnesota and natural resource management issues, see McClurken et al., *Fish in the Lakes.* This book greatly helped me understand why land use and Native autonomy issues are deep rooted and not easily resolved. I highly recommend this book to anyone working on treaty rights issues.
5. Hotchkiss, *Lumber and Forest Industry*, 641.
6. Roberts, *Trending into Maine*; Morison, *Maritime History of Massachusetts.*
7. Barton, White, and Cogbill, *Changing Nature*, 115.
8. Barton, White, and Cogbill, 116, figure 4.11.
9. Thoreau, *Maine Woods*, 133.
10. Baskerville, "Degrading Forest System," 314–22.
11. Barton, White, and Cogbill, *Changing Nature*, 116, figure 4.12.
12. Baskerville, "Degrading Forest System," 314–22.
13. Hemingway, *The Nick Adams Stories.*
14. Abrams, "Eastern White Pine Versatility," 967–79.
15. Hotchkiss, *Lumber and Forest Industry*, 638.
16. Larson, *White Pine Industry*, 133.
17. Quoted in Rutkow, *American Canopy*, 111.
18. Holbrook, *Holy Old Mackinaw.*
19. Hotchkiss, *Lumber and Forest Industry*, 749.
20. For a history of the Hinckley Fire, see Brown, *Under a Flaming Sky.* The stories of other fires are essentially the same.

Chapter 4. Thoreau, the Maine Woods, Forest Succession, and Faith in a Seed

1. Thoreau, *The Maine Woods*, 112.
2. Thoreau.
3. Detailed maps, itineraries, and notes on the natural history of Thoreau's three Maine trips can be found in Huber, *The Wildest Country*.
4. Thoreau, *The Maine Woods*, 13.
5. Thoreau, 15.
6. Thoreau, 17.
7. A fish hawk is an osprey (*Pandion haliaetus*).
8. Thoreau, *The Maine Woods*, 73.
9. Thoreau, 133.
10. Thoreau, 100.
11. Today we would probably say *compost* instead of manure.
12. Thoreau, *The Maine Woods*, 112.
13. Cicero, *On the Republic. On the Laws*, 333; see also Thoreau, *The Maine Woods*, 112n98.
14. Walls, "'As Planets Faithful Be,'" 171–94.
15. Thoreau, *The Maine Woods*, 112.
16. Thoreau, 118; Thoreau, *Walden*, 39.
17. Thoreau, *The Maine Woods*, 140.
18. Thoreau, 141.
19. Thoreau, 144.
20. There seems to be some confusion about Thoreau's family relationship to Charles Lowell. Walls (*Henry David Thoreau*, 219) says he is Thoreau's brother-in-law, but in 1906, eds. Torrey and Allen (*Journal of Henry D. Thoreau*) claimed Lowell was a cousin. In his journal, Thoreau just calls him Lowell.
21. *Journal of Henry D. Thoreau*, "September 22, 1853," vols. 1–7, 432–33.
22. Quoted in Thoreau, *The Maine Woods*, 198.
23. Thoreau, 195.
24. Thoreau, 213.

25. Thoreau, 235.
26. *Journal of Henry D. Thoreau*, "March 9, 1855," vols. 1–7, 854.
27. Thoreau, *The Natural History Essays*, 91.
28. Dean, "Thoreau and Horace Greeley," 630–38.
29. *Journal of Henry D. Thoreau*, "September 12, 1857," vols. 8–14, 1190.
30. Thoreau, *The Natural History Essays*, 84.
31. Thoreau, *Faith in a Seed*, 34.
32. This seemingly odd unit is one-quarter of a surveyor's chain (sixty-six feet), literally a chain of that length that surveyors once used to mark distances for property surveys. The reason for these oddly sized units is that an acre can be surveyed using even numbers of rods or chains. A "perfect acre" is a rectangular plot of land forty rods (ten chains) on the long side and four rods (one chain) on the short side; that is, 160 square rods, or forty square chains. Thoreau made a living for much of his life by surveying property lines; much of this land would be sold to loggers, an irony not lost on him. Thoreau was highly regarded as an accurate and honest surveyor not only in Massachusetts but even as far away as New Jersey. Thoreau's journals are full of natural history observations made while he surveyed various properties. For more about Thoreau's life as a surveyor, see Chura, *Thoreau the Land Surveyor*.
33. Thoreau, *Faith in a Seed*, 35.
34. Harper (*Population Biology of Plants*, 609) makes the analogy between the invasion of abandoned farm fields by pines and the spread of epidemics. I happen to be writing this during the 2020 coronavirus pandemic, which spread in a similar manner.
35. For Hubbard's remarks, see *Journal of Henry D. Thoreau*, "April 28, 1856," vols. 8–14, 1109; for confirmation of what Hubbard told him, see Thoreau's journal entries dated May 13 and June 1, 1856, vols. 8–14, 1014, 1021.
36. *Journal of Henry D. Thoreau*, "October 14, 1860," vols. 8–14, 1895.
37. Thoreau, *Faith in a Seed*, 121.
38. Thoreau, *The Natural History Essays*, 78.
39. Thoreau, *Faith in a Seed*, 106.
40. *Journal of Henry D. Thoreau*, "November 24, 1860," vols. 8–14, 1730.
41. Clements, *Dynamics of Vegetation*.

42. Wein and El-Bayoumi, "Predictability of Plant Succession," 214–25.
43. Walls, *Seeing New Worlds*.

Chapter 5. The Watershed

1. Carlson, Munroe, and Hegman, "Distribution of Alpine Tundra," 331–42.
2. Keller, *Adirondack Wilderness*.
3. For an excellent image and commentary on the painting, see Terra Foundation, *Conversations with the Collection: A Terra Foundation Collection Handbook*, "Landscape Painting: Sanford Robinson Gifford (1823–1880)," https://conversations.terraamericanart.org/artworks/hunter-mountain-twilight.
4. Johnson, "From a Woodland Elegy."
5. Marsh, *Artificial Propagation of Fish*, 7–9; see also Trombulak, *So Great a Vision*, 67–68.
6. Stegner, *Where the Bluebird Sings*, 123.
7. Marsh, *Man and Nature*, 160n, 222.
8. I think I heard Bill Reiners, a distinguished forest ecologist first at Dartmouth College and then at the University of Wyoming, say this at an Annual Meeting of the Ecological Society of America, but I can't remember when or where it was, so I contacted Reiners and asked him. He couldn't remember where he'd said it either, but he wrote to me that it sounded like things he has said over the years, most especially in his 2004 book with Ken Dreise, *Transport Processes in Nature*.
9. Marsh, *Man and Nature*, 27.
10. Marsh, 204.
11. Keller, *Adirondack Wilderness*; Jacoby, "Class and Environmental History," 324–42.
12. The definitive biography of Gifford Pinchot is Miller, *Gifford Pinchot and the Making of Modern Environmentalism*.
13. Pinchot, *Breaking New Ground*, xvi–xvii.
14. Pinchot, 73.
15. Pinchot, *Fishing Talk*, 233.
16. Brinkley, *Wilderness Warrior*; Canfield, *Roosevelt in the Field*.

17. Swank and Miner, "Conversion of Hardwood-Covered Watersheds," 947–54.
18. Helvey, "Interception," 723–29.
19. Swank and Miner, "Conversion of Hardwood-Covered Watersheds," 947–54; Swift et al., "Evapotranspiration and Drainage," 667–73.
20. Demuth, *Floating Coast*, 126.

Chapter 6. A Scientific Foundation of White Pine Ecology and Management

1. Still the best economic analysis of this dilemma is by the great economist Paul Samuelson ("Economics of Forestry," 466–92).
2. Peckham, *University of Michigan*, 86.
3. Peckham.
4. Gray, *Manual of the Botany*.
5. Spalding, "Progress of Ecology," 201–10.
6. Reed, "Volney Morgan Spalding," 14–18.
7. Alas, the administrative duties that draw faculty from research and teaching were no less at that time than they are today.
8. The standard biography of Fernow is Rodgers, *Bernard Eduard Fernow*.
9. Pinchot and Graves, *The White Pine*.
10. Spalding and Fernow, *The White Pine*.
11. Burns and Honkala, *Silvics of North America*.
12. Along with the diminutive and beautiful blue and copper butterflies, the elfins are in the family Lycaenidae, the second-largest family of butterflies. The Lycaenidae were studied most notably by the author Vladimir Nabokov, who was also a respected lepidopterist; see Johnson and Coates, *Nabokov's Blues*.
13. At half past four in the afternoon on May 28, 2020, a sunny and pleasant day, I saw an eastern pine elfin nectaring in a small patch of wild strawberries in my front yard within a few hundred yards of a large stand of old-growth white pine, along with the closely related Henry's elfin (*Callophrys henrici*) and two slightly more distantly related lycaenids, the spring azure (*Celastrina ladon*) and the silvery blue (*Glaucopsyche lygdamus*). A feast of lycaenid butterflies.

14. Pinchot and Graves, *The White Pine.*
15. The reasons why trees and forests grow more slowly as they age are surprisingly difficult to pin down. For a comprehensive review of a number of hypothesized mechanisms, see Ryan, Binkley, and Fownes, "Decline in Forest Productivity," 213–62.
16. Kershaw et al., *Forest Mensuration.*
17. Grinnell, *Philosophy of Nature.*

Chapter 7. Rusty Pines and Gooseberries

1. The white pine is still today called Weymouth pine in Great Britain, although whether that name came from George or Thomas is debated. The irony of Viscount Weymouth's planting scheme is that by the time his white pine seedlings grew to the size needed for ship's masts, the end of the days of sail would be rising over the horizon.
2. Spaulding, *Blister Rust.*
3. For an excellent theoretical treatment of why diseases need multiple hosts to persist, see Dobson, "Population Dynamics of Pathogens," S64–S78. Although Dobson relies on examples of animal diseases, his theoretical model applies equally well to plants.
4. For detailed descriptions of the life cycle of blister rust, see Spaulding, *Blister Rust*; Maloy, "Blister Rust Control," 87–109; and Geils, Hummer, and Hunt, "White Pines, *Ribes*," 147–85.
5. *Ribes* is the Latin name for the genus, but it is also widely used as the common name for all the gooseberries and currants together. I will use it in this sense throughout this chapter.
6. The evolution of this subgenus is discussed in chapter 1.
7. These spores are haploid, unlike the other spores that are diploid. For more details, see Maloy, "Blister Rust Control," 87–109; and Geils, Hummer, and Hunt, "White Pines, *Ribes*," 147–85.
8. Van Arsdel, "Nocturnal Diffusion and Transport," 1221–29; Robbins, Jackson, and McRoberts, "Eastern Upper Peninsula," 263–64; Dahir and Carlson, "High-Hazard Region of Wisconsin," 81–86, figure 5.
9. Kalm, *Travels in North America.*
10. A footnote on the title page states: "This Act must be cited as the Act of August 20, 1912 (37 Stat. 315, chapter 308; 7 USC. 151 et seq.)." It is

only commonly known as the Plant Quarantine Act or the Nursery Stock Quarantine Act.

11. Maloy, "Blister Rust Control," 87–109.
12. Spaulding, *Blister Rust.*
13. Spaulding, *Investigations.*
14. For details of the various programs to eradicate blister rust, see Maloy, "Blister Rust Control," 87–109.
15. An incident that is famous in the annals of blister rust control was the so-called Battle of Kittery Point. Professor Roland Thaxter, a distinguished mycologist at the University of Connecticut Agricultural Experiment Station, had a summer place in Kittery Point in the southernmost corner of Maine, where he grew currants and gooseberries in his garden for jams and winter preserves. When federal blister rust agents showed up to advise him to destroy these plants under penalty of law, Thaxter met them at the gate with a shotgun. Horsfall, "Roland Thaxter," 29–35.
16. Maher, *Nature's New Deal,* 56.
17. On July 14, 2021, while we were counting northern blue and other butterflies at McNair, Minnesota, site of a former logging camp and near some extensive CCC red pine plantations.
18. King, *Lake States.*
19. Maloy, "Blister Rust Control," 87–109.
20. Maloy.
21. Bonner, *Life Cycles.*

Chapter 8. Roosevelt's Tree Army

1. In telling the story of John Ripley and the opening of Camp Roosevelt, various sources say that Ripley climbed a "pine" tree. I have not been able to confirm what species of pine it was, but of all the pine species in the Shenandoah Valley, white pine would have been the sturdiest and tallest. In addition, a photo on the Wikipedia page for Civilian Conservation Corps shows CCCs at Camp Roosevelt eating a meal outside on picnic tables with what appears to be some white pines in the background. Kauffman, "Roosevelt—Forest Camp," 251–54; Maher, "New Deal Body Politic, 435–61; Maher, *Nature's New Deal.*
2. Nowak, *Cultural Landscape Report.*

3. Brinkley, *Righful Heritage*, 65–67.
4. Maher, *Nature's New Deal*, 50.
5. Quoted in Miller, *Gifford Pinchot*, 287.
6. Benson, *Stories in Log and Stone*. The enrollees' nickname for their organization was Colossal College of Calluses, which pays tribute to not only the hard physical work but also the pride the men took in doing it.
7. Brinkley, appendix F of *Rightful Heritage*, 625–26.
8. Brinkley, *Rightful Heritage*, 182. The young men from rural and small-town America included several of my and my wife's uncles.
9. Brinkley, appendix F of *Rightful Heritage*, 625.
10. To put this in some perspective, in the years before the CCC was established, only four hundred acres of white pine were planted annually in Wisconsin, which was one of the more active states in forest restoration. Gevorkiantz and Zon, *Second-Growth White Pine*.
11. During World War II, when the young men of the CCC were fighting in the European and Pacific theaters, these towers were staffed by young women from local towns. An interview with one of these women appears in Irvine, "Women Manned the Mountain," 54–57.
12. The data in this paragraph derived from McEntee, "Final Report."
13. Maher, *Nature's New Deal*, 100, 102.
14. For an interesting history of the Black Rock Forest and the role it played in the development of American forestry and forest ecology, see Trow, *Harvard Black Rock Forest*.
15. Mitchell, *Nutrition of White Pine*.
16. Mitchell, Finn, and Rosendahl, "Relation between Mychorrhizae" 58–73; Hacskaylo and Snow, *Soils Nutrients and Light*.
17. Maser, Trappe, and Nussbaum, "Fungal–Small Mammal Relationships," 799–809; Pastor, Dewey, and Christian, "Carbon and Nutrient Mineralization," 52–61.
18. Terwilliger and Pastor, "Small Mammals, Ectomycorrhizae," 83–94.
19. Klironomos and Hart, "Animal Nitrogen Swap," 651–52.
20. For an excellent review of the role of mycorrhizae in ecosystems and in reforestation, see Perry, Molina, and Amaranthus, "Mycorrhizae,

Mycorrhizospheres, and Reforestation," 929–940. See also Simard, *Finding the Mother Tree.*

21. Foresters say the sapling stage begins when the white pine is about five inches in diameter at breast height (4.5 feet above the forest floor) and continues until the tree is about fifteen to twenty years old, when it begins to produce cones.
22. Porter, "African Americans in the CCC."

Chapter 9. Rebirth by Fire

1. Minor and Boyce, "Smokey Bear," 79–93.
2. Rice, "Smokey Bear," 276–79.
3. Davidson-Hunt, "Indigenous Land Management," 21–42; Berkes and Davidson-Hunt, "Biodiversity, Traditional Management Systems," 35–47. Ironically, foresters in the Chequamegon National Forest in northern Wisconsin today maintain a large expanse of blueberries in the Moquah pine barrens by periodic prescribed fires.
4. Russell, "Indian-Set Fires," 78–88.
5. Heinselman, *Boundary Waters Wilderness Ecosystem*, xi.
6. Heinselman, "Forest Sites, Bog Processes," 327–74; "Landscape Evolution, Peatland Types," 235–61.
7. Wilderness Act of 1964, Pub. L. No. 88-577, 16 U.S.C. § 1131–36 (1964).
8. See especially chapter 10 in Johnson and Govatski, *Forests for the People.*
9. The Quetico land has also been designated a wilderness area by the Canadian government and is now known as Quetico Provincial Park. The US Forest Service and the Ontario Ministry of Natural Resources coordinate efforts in management of the entire Quetico–Superior.
10. For more on the legal history of the Boundary Waters Wilderness, see Proescholdt, Rapson, and Heinselman, *Troubled Waters*; Proescholdt, "First Fight," 70–84.
11. Quoted in Heinselman et al., "Wilderness Biotic Communities."
12. Heinselman et al.
13. Miron L. Heinselman Papers, Minnesota Historical Society Library, Location 141.C.7.5B, box 2.
14. Clements, *Lodgepole Burn Forests.*

15. Clements, "Structure of the Climax," 252–84.
16. For a discussion of the evolution and ecology of serotiny, see Pastor, *A Clever Moose*, chap. 29.
17. Digitized images of these maps can be found at the University of Minnesota Libraries Digital Conservancy, Center for Forest Ecology, maintained by the University of Minnesota John R. Borchert Map Library, https://conservancy.umn.edu/handle/11299/168076/.
18. These trees, three red pines on Three Mile Island in Lake Saganaga that we visited on this field trip, are no longer standing, having been toppled in several windstorms. You can see a photo, taken during this field trip, of Bud Heinselman pointing to the hole where he took the core from one of these pines on the inside flap of the dust jacket of Heinselman, *Boundary Waters Wilderness Ecosystem*.
19. Heinselman, "Fire in the Virgin Forests," 379–82.
20. Heinselman.
21. Heinselman; see also Heinselman, "Fire Intensity and Frequency," 7–57.
22. Heinselman, "Fire in the Virgin Forests," 379–82; Frelich and Reich, "Neighborhood Effects, Disturbance," 148–58.
23. Lopez, *Arctic Dreams*, 25.
24. Loucks, "Evolution of Diversity, Efficiency," 17–25.
25. US Forest Service, *Fire Management Action Plan*.
26. This fire policy and its caveats are discussed in detail in Heinselman, *Boundary Waters Wilderness Ecosystem*, 144–59.
27. See Pastor, *A Clever Moose*, chap. 19, for a description and discussion of the ecology and Forest Service policy regarding the Pagami Creek Fire in the Boundary Waters, a very large natural fire that burned in 2011 and which resembled many of the historic fires documented by Heinselman, "Fire in the Virgin Forests."
28. Leopold, "Land Laboratory," 287–89.

Chapter 10. Restoring the White Pine

1. Information on the Rajala family and their role in white pine logging in northern Minnesota is taken from Rajala, *Tim-BERR!*; Rajala, *White Pine*; the website of the Rajala companies, https://mntimber.com/; and conversations with John Rajala in the forests of the Wolf Lake Tract.

2. Rajala, *White Pine.*
3. Puettmann, Coates, and Messier, *Critique of Silviculture.*
4. Palik et al., *Ecological Silviculture*, xiii.
5. Tester, Starfield, and Frelich, "Modeling for Ecosystem Management," 313.
6. Fahey and Lorimer, "Restoring a Midtolerant Pine," 139–49.
7. Weyenberg, Frelich, and Reich, "Logging versus Fire," 179–94.
8. White, "Effects of Deer Browsing," 222–28.
9. Nuttle, Ristau, and Royo, "Long-Term Biological Legacies," 221–28.
10. See Rajala, "Why We Bud Cap"; Rajala, "Bud Capping," YouTube video.
11. For an example of the decisions that have to be made on the ground, see Rajala, "Northern Red Oak Shelterwood," YouTube video.
12. Much of what is sold as "white pine" in big-box lumberyards is not eastern white pine (*Pinus strobus*) but *Pinus radiata*, otherwise called radiata pine or Monterey pine. Radiata pine is native to California but has become the most widely planted pine in the world, especially across the Southern Hemisphere, which has no native pine species. Ninety-five percent of timber produced in Chile comes from radiata pine plantations, which have often displaced native forests. The irony is that, although the wood has a light color similar to eastern white pine, radiata pine is a member of the yellow pine subgenus of *Pinus* rather than the white pine subgenus for which eastern white pine is the type species.
13. Trosper, "Indigenous Influence," 134–39.
14. For case studies of ecological silviculture in various National Forests, see Palik et al., *Ecological Silviculture.*
15. Harmon, Ferrell, and Franklin, "Effects on Carbon Storage," 699–702; Dewar, "Model of Carbon Storage," 239–58; Dewar, "Carbon Storage in Forests," 417–28.
16. Fargione et al., "Natural Climate Solutions."

Chapter 11. Climate Change and the Future of White Pine

1. Seltzer et al., "Six Degrees Celsius Cooling," 228–32.
2. For changes in the distribution of white pine in response to various climate forecasts, see USDA Forest Service, "Tree Atlas, version 4."

3. To name just a few of many: Pastor and Post, "Response of Northern Forests," 55–58; Overpeck, Bartlein, and Webb, "Future Vegetation Change," 692–95; Iverson and Prasad, "Predicting Abundance," 465–85.
4. USDA Forest Service, "Tree Atlas, version 4."
5. Davis and Shaw, "Range Shifts," 673–79.
6. Davis, Shaw, and Etterson, "Evolutionary Responses," 1704–14.
7. Etterson and Shaw, "Constraint to Adaptive Evolution," 151–54.
8. Hoegh-Guldberg et al., "Assisted Colonization," 345–46; Winder, Nelson, and Beardmore, "Implications for Assisted Migration," 731–44.
9. Pedlar et al., "Placing Forestry," 835–42.
10. Etterson et al., "Assisted Migration," e02092.
11. McLachlan, Hellman, and Schwartz, "A Framework for Debate," 297–302; Ricciardi and Simberloff, "Assisted Colonization," 248–53; Aubin et al., "Why We Disagree," 755–65.
12. Hunter, "Climate Change," 1356–58; Galatowitsch, Frelich, and Phillips-Mao, "Climate Change Adaptation Strategies," 2012–22.
13. Millar, Stephenson, and Stephens, "Forests of the Future," 2145–51.
14. Anderson, Clark, and Sheldon, "Estimating Climate Resilience," 959–70.
15. Dobrowski, "Climatic Basis for Microrefugia," 1022–35.
16. White et al., "Take on Climate Change," 219–32.

Afterword

1. Pyle, "Rise and Fall," 235.

About the Author

MARY DRAGICH

John Pastor is an ecologist and professor emeritus at the University of Minnesota Duluth, where his teaching and research focused on the natural history and ecology of northern ecosystems. He is the author of *What Should a Clever Moose Eat? Natural History, Ecology, and the North Woods* and *Mathematical Ecology of Populations and Ecosystems*, and coeditor of *Large Mammalian Herbivores, Ecosystem Dynamics and Conservation*. Pastor has authored or coauthored twenty-two book chapters and more than 120 papers, mostly about the North Woods. He is a past cochair of the Natural History Section of the Ecological Society of America and founding editor of the Scientific Naturalist series in the journal *Ecology*.

Index